AF775703

Healthcare 5.0

Kamboj Dalip · Urvashi Bansal · Arvind Sharma ·
Samiya Khan · Monika Sharma
Editors

Healthcare 5.0

AI Driven Workspace in Sustainable Telehealth

Springer

Editors
Kamboj Dalip
M. M. Institute of Computer Technology
and Business Management
M.M (Deemed to be University)
Mullana, Haryana, India

Arvind Sharma
Yogananda School of AI
Computers and Data Science
Shoolini University
Solan, Himachal Pradesh, India

Monika Sharma
Mobile Medical Unit
National Health Mission
Solan, Himachal Pradesh, India

Urvashi Bansal
Department of Computer Science
and Engineering
Dr. B. R. Ambedkar National Institute
of Technology Jalandhar
Jalandhar, Punjab, India

Samiya Khan
School of Electronics and Computer
Science
University of Southampton
Delhi, India

ISBN 978-3-032-09581-7 ISBN 978-3-032-09582-4 (eBook)
https://doi.org/10.1007/978-3-032-09582-4

© The Editor(s) (if applicable) and The Author(s), under exclusive license to Springer Nature Switzerland AG 2026

This work is subject to copyright. All rights are solely and exclusively licensed by the Publisher, whether the whole or part of the material is concerned, specifically the rights of translation, reprinting, reuse of illustrations, recitation, broadcasting, reproduction on microfilms or in any other physical way, and transmission or information storage and retrieval, electronic adaptation, computer software, or by similar or dissimilar methodology now known or hereafter developed.
The use of general descriptive names, registered names, trademarks, service marks, etc. in this publication does not imply, even in the absence of a specific statement, that such names are exempt from the relevant protective laws and regulations and therefore free for general use.
The publisher, the authors and the editors are safe to assume that the advice and information in this book are believed to be true and accurate at the date of publication. Neither the publisher nor the authors or the editors give a warranty, expressed or implied, with respect to the material contained herein or for any errors or omissions that may have been made. The publisher remains neutral with regard to jurisdictional claims in published maps and institutional affiliations.

This Springer imprint is published by the registered company Springer Nature Switzerland AG
The registered company address is: Gewerbestrasse 11, 6330 Cham, Switzerland

If disposing of this product, please recycle the paper.

Contents

Chapter 1
The Role of Artificial Intelligence in Chronic Disease Management Through Telemedicine and Future Directions

Wasswa Shafik

Abstract Artificial intelligence (AI) research is a vital area that flourishes in many fields, including healthcare. Chronic diseases are increasing at an alarming pace, which reduces the quality of life as well as economic growth. Telemedicine is a technique to provide healthcare facilities to remote areas or medically underserved populations by making use of technology. Healthcare costs and non-compliance with medication have further reduced patient adherence to their daily self-care for chronic management; AI plays a significant role in reducing healthcare costs and improving patient care. Over 2.5 quintillion bytes of data are created every day. It is an impossible task to utilize this large amount of patient data effectively for a healthcare professional in making a complex drug prescription treatment for a chronic disease. Thus, the traditional way of pharmaceutical care to address chronic diseases may not be sustainable in the future due to the rise of patients with multiple drug regimens and increasing costs of medications. In the future, the AI system will take its place: "Robots" will see you now. "Telemedicine has ultimately transformed the working standard of any country's healthcare system, and many countries have accepted telemedicine as the latest means of healthcare facilities. Virtual care adoption increased significantly for video, phone, and messages. There are only 13 states with a surplus in telemedicine as of 2030. In recent years, significant advancements have occurred in technology, leading to minimally invasive surgical interventions with better patient post-surgery life quality and better monitoring and screening techniques for chronic patients. The AI system uses algorithms in decision-making for telemedicine and clinical care, which reduces healthcare costs and helps patients take medication at the correct dosage and time with better adherence. The healthcare system now uses AI for image analysis, diagnostic decision support, virtual clinical assistants, and pattern analysis to monitor patient data and provide recommendations on clinical treatment. The AI engine can be either handheld devices or software

W. Shafik (✉)
School of Digital Science, Universiti Brunei Darussalam, Jalan Tungku Link, Gadong, Bandar Seri Begawan, Brunei Darussalam, Brunei
e-mail: wasswashafik@ieee.org

© The Author(s), under exclusive license to Springer Nature Switzerland AG 2026

K. Dalip et al. (eds.), *Healthcare 5.0*, https://doi.org/10.1007/978-3-032-09582-4_1

through the internet. This study discusses the role of AI in telemedicine for chronic disease.

Keywords Artificial intelligence · Chronic disease management · Telemedicine · Predictive analytics · Remote patient monitoring · Health innovation · Personalized treatment · Future healthcare trends

1.1 Introduction

The use of imaging and diagnostics through machine learning, robotics, and remote operational technologies has opened the window to modern medicine. Despite the rapid development of high-tech diagnostic systems, they have not made significant strides in curing chronic diseases. The rapidly evolving machine learning models have unlocked a set of complex algorithms to evaluate a plethora of terabytes of data and are integrating AI into modern telemedicine, making it more potent to diagnose chronic diseases like cancer, diabetes, and heart failure [1]. Discuss the complex interaction between AI and computational aspects involved in chronic disease management through telemedicine. Subsequently, we discuss current problems in telemedicine, however, we also outline prospects in chronic disease management with AI-driven advancements. The incidence of chronic diseases is increasing at an alarming rate, posing a significant healthcare concern worldwide. The burden on the healthcare system is enormous, as chronic diseases account for over 50% of the disease burden and an estimated 60% of healthcare expenses across the globe [2]. AI is anticipated to revolutionize patient care by providing more precise and accurate data-driven insights and thus improve diagnostic and therapeutic patient management. Thus, this paper aims to briefly articulate the current role of AI in telemedicine for chronic disease management and discuss prospects. The purpose of this paper was to explore the current roles of AI in telemedicine for chronic diseases and to provide the rationale for future research in the field.

Chronic diseases differ from acute illnesses based on their inability to resolve spontaneously in a short period and their tendency to persist over a longer duration. There's consensus on a 3 to 6-month minimum timeline for a condition to be classified as chronic. Acute diseases, by definition, have a short course, a variety of etiologies, and discrete outcomes ranging from complete recovery to death. They are also often broadly classified as diseases of prevention, but chronic conditions are generally more focused upon. Chronic diseases can have permanent functional or anatomical damage, condition-induced disability, higher rates of co-morbid conditions, and lethality, especially when left uncontrolled [3, 4]. Multifactorial etiologies underpin chronic diseases, and treatment planning is additionally complicated by co-morbidity on multiple levels. Chronic diseases represent long-term stressors and challenges that require ameliorative strategies and a lifestyle management approach rather than having a sharp reality that quickly drives the host into forbidden paths. Behavioral dynamics often underpin disease etiology, and risk factor modifications are critical focal points of their management. At present, 7 out of 10 deaths in the United States occur due to chronic diseases, and a considerable percentage of healthcare dollar spending is directed towards their care [5]. In addition to the direct morbidity and mortality resulting from chronic diseases themselves, unplanned visits to the hospital or treatment center also often result from the decompensation, exacerbation, or progression of chronic co-morbid conditions. It is often stated that chronic care has become the mainstay of healthcare today. Public health initiatives currently focus on the identification of determinants and risk factors for the development of chronic disease, as well as early intervention and prevention. The care and maintenance of individuals diagnosed with chronic disease is additionally a particularly challenging aspect of this new focal point in health [6].

A physiological, psychiatric, social, or functional condition persisting for a prolonged period, usually for more than a year, is considered a chronic condition. There is no consensus on categorization, but the designation is made based on certain features, such as the duration of the disease, the progression of the disease stage and associated complications, irreversible injuries caused by the disease or sequelae, and impairment in daily living [7]. Conditions defined as "chronic" impair confidence and self-reliance to live their lives. Treatment and follow-up of the diseases can be provided in the same polyclinic, hospital, or outpatient and inpatient clinics in which the diagnosis is made. Infectious diseases, on the other hand, if kept under control, usually terminate within a short period unless complications arise [8]. However, long-term treatment is occasionally necessary for the effective eradication or control of illness or to achieve complete protection.

(1) Comorbidity: Depending on full-scale data, it has been previously reported that between 53 and 95% of chronic patients suffer from more than one chronic disease at the same time. One of the main priorities of healthcare systems is to improve the management of patients suffering from chronic diseases, mainly depending on the type, such as classification of communication that does not possess at least one academic degree in medicine and healthcare providing divisions and topics like classification of healthcare providing services and divisions, not in terms of education-based.
(2) Some "global epidemics" originated from chronic diseases. Chronic diseases can be triggered by factors such as smoking, obesity, stress, and so on, but since chronic diseases do not spread from person to person, their uncontrolled continuation over time will not cause new chronic diseases.

The combination of two or more chronic diseases or the development of one type of chronic disease in another chronic disease patient simultaneously is defined as "comorbid" or "multimorbid." Comorbidity can fail one disease treatment while complicating the progression of the other disease. Based on the criteria, some well-known diseases are chronic, such as simple goiter, voiding disorders, mild asthma, mild gastroesophageal reflux disease, type 2 diabetes mellitus, mild hypertension, mild heart failure with normal left ventricular ejection fraction, and impairment in daily living is less common than type 1 diabetes, pulmonary embolism sequelae due to primary vascular disease with average plasma sodium concentration, mild *hypokalemia*, moderate chronic *pulmonary* chronic obstructive pulmonary disease (COPD), but comorbid mild chronic muscular depression, stage 3 chronic renal failure, mild diabetic retinopathy, hypertensive heart disease associated with angina pectoris, and symptomatic chronic atrial fibrillation accompany chronic hyperthyroidism during warfarin treatment [9]. In the event that patients contract these diseases at the same time, they will be considered comorbid and will both be treated accordingly [10]. In this case, because each patient has three chronic diseases at the same time, the patient falls under the category of multimorbidity. In practice, the probability of developing a chronic disease increases with age, and those over 70 years are more likely to develop multimorbidity.

The term chronic disease refers to a health problem requiring long-term management, is progressive, has permanent states, or is incurable. Chronic diseases are non-communicable diseases, usually associated with many of the same lifestyle choices as mental health issues resulting from the current epidemic of these conditions. Moreover, the aging population in China is reported to now have just under 5% with a chronic disease, while over 25% are currently affected in China and the United States. Globally, they are the leading causes of disability and mortality, with an estimated 41 million deaths from chronic disease in 2016, divided up as cardiovascular, cancers, chronic respiratory, and diabetes, with an additional from all mental health disorders. While these rates of deaths by incidence may seem exceptionally high, the disease's actual prevalence is even higher [11, 12]. Chronic disease now affects 8.2% of children and 10.3% of the adult populations worldwide. Interestingly, 48% of those with one chronic disease have a co-existing second, and another almost 30% have 3 or more long-term conditions. Several international studies document that the global burden of chronic diseases varies by age, sex, and socioeconomic status, with older women who are generally sicker; conversely, more men die from the condition than women. These diseases are also not race-free. In Australia, more than 70% of Aboriginal and Torres Strait Islander people are already living with a chronic disease, and by age 35, 100% will receive a medical diagnosis for a chronic condition, compared with 70% of non-Indigenous people [6]. This information suggests an urgent need for programs to prevent and better manage these long-term health conditions.

Chronic diseases in recent days are on the rise due to various factors like aging, changes in lifestyles, and poor biometrical conditions. This rising number of chronic conditions leads to another problem of managing chronic diseases and places a burden on the healthcare systems. Although much research is being done to prevent the onset or eradication of these diseases, the number of chronic conditions and the individuals being diagnosed each year has not decreased significantly over the past decades. The traditional approach of the healthcare system to provide solutions only after the diagnosis is not a cost-efficient one in the case of chronic diseases. Having realized the importance of chronic disease management to avoid the resultant complexities, there have been several research efforts to develop efficient strategies to monitor the condition of chronic and abnormal populations effectively, one such with the promise of incorporating telemedicine and the service model of healthcare for the same [5]. One of the potential solutions to the challenges pertaining to the area of chronic disease management has been to blend innovations from other sectors and leverage the same to overcome the challenges. In the progress of the health sector, healthcare technology has played a significant role, and the technology has dramatically improved over the past centuries. In the early stages, the invention of technologies was mainly focused on the diagnosis of patients, and the available diagnostic technologies may not be linked to an existing system to generate a cumulative report of the population to monitor the health behavior of an area. The world of healthcare entered a new era with several modern inventions like electronic health record systems, communication systems that could be used to facilitate healthcare management over a remote area with its people, real-time patient monitoring devices, and wearables. These kinds of healthcare products and services are used to treat individuals who are healthy and abnormal, irrespective of the range [13, 14]. Note that telemedicine has its role and has become very popular nowadays as this type of healthcare is directly conveyed involving the patients and the professionals and enables the validation of the diagnosis if required. In recent telemedicine approaches, the majority of the use has been to deal with individuals affected by disorders.

1.1.1 The Chapter Contribution

This chapter presents the following contributions:

- This chapter explores and analyzes what is potentially the most crucial issue facing the world of medicine in the twenty-first century, particularly from the standpoint of patient-based care. Therefore, it is essential to delimit or indicate the parameters of our discussion.
- This confines the discussion to consider AI in relation to telemedicine generally and in a variety of possible scenarios related to chronic disease management. It is designed to help healthcare providers and systems, policymakers, and researchers who may be interested in technology-facilitated services currently or in the future.
- The research searches, identifies and explores in an in-depth manner the potential of AI in already operative telemedicine cases and, in the near future, to identify this technique in expert systems for the automated management of chronic diseases.
- The chapter presents a brief overview of each section, and its need for inclusion in this work is now introduced, entailing an outline of the way in which the content of this chapter unfolds, looking first at the benefits of AI in medicine and telemedicine and then examining some of the barriers or impediments to its adoption.
- Like telemedicine, most studies of AI in telemedicine focus on chronic diseases or, more specifically, on cancer and home care. To date, no comprehensive review of the application of expert systems in chronic disease management and its implications has been done, as this chapter presents.
- Incredibly, despite rampant use by more in developed nations, developing countries remain ignorant of the health potential of AI. Managerial implications of

expecting systematic transformation through the use of AI in healthcare are discussed throughout the chapter.
- Finally, within this chapter, two contrasting case studies are presented and explained in detail on how they are applied in the healthcare domain, followed by practical recommendations.

1.1.2 The Chapter Organization

The remainder of this study is structured as follows. Section 1.2 presents the Artificial Intelligence in Healthcare, depicting the overview of artificial intelligence and its application. Section 1.3 presents telemedicine in chronic disease management, detailing an understanding of the scope of telemedicine, telemedicine benefits, and challenges. Section 1.4 presents the integration of AI in telemedicine for chronic disease management, including current status and trends, key technologies, and approaches. Section 1.5 illustrated sampled case studies and success stories of its application, including hypertension, diabetes, clinic pilot programming, and impacts on patient outcomes. Section 1.6 presents telemedicine's future directions and emerging trends entailing advances in AI and telemedicine, potential benefits and risks. Section 1.7 details ethical and legal considerations such as data privacy and security, bias, and fairness in AI, among others. Section 1.8 presents recommendations presenting a summary of key findings and Implications for Healthcare practitioners. Finally, Sect. 1.9 presents the chapter's conclusion.

1.2 Artificial Intelligence in Healthcare

AI refers to advanced technologies that allow computer systems to perform tasks typically requiring human intelligence, such as visual and speech recognition, language translation, and decision-making. AI technologies can be broken down into three key components: natural language processing (NLP), machine learning (ML), and deep learning (DL). For use in healthcare, AI can integrate functionalities of the three core components where it can compare, translate, process, and learn healthcare data up to par with data analytics. AI systems can support health informatics by accurately processing big data from different dimensions in a more efficient manner when compared to statistical tools, lowering time-consuming human efforts in terms of analysis [10].

AI technologies come with algorithms such as Random Forest, Gradient Boosting Machine, k-Nearest Neighbor, Naive Bayes, Support Vector Machine, and Artificial Neural Networks, being the popular ones that can be tailored to respective use cases in the healthcare system. Using predictive analytics and pattern identification functions in medical image diagnosis, drug discovery, triage, automated treatment plans, assisting physicians in interpreting lab test results, and early detection of chronic diseases. AI can also advance personalized medicine and have direct applications in telemedicine. AI's paramount importance is its capability to enhance the efficiency at which data results are accessed and custom-tailored for personalized treatment options, improving precision diagnostics and mitigating errors and patient safety risks [2, 15]. The role of AI technologies in chronic disease management is extensively discussed. AI would develop diagnostic devices and smart primary care services through a personal health record and lifestyle management portal. Electrocardiogram monitoring technologies are also under development for long-term monitoring of the electrocardiogram and event recording of arrhythmias using early detection and risk stratification algorithms. The most promising tool to improve telemedicine is integrated AI models that bridge predictive technologies with high-quality diagnostic data. Such systems improve the predictive power of AI models by adding real-time, time-stamped patient data [16].

They allow AI systems to dynamically update and refine predictions as each patient result comes to light. Telemedicine is another ripe area for incorporating AI. AI could steer triaging systems for patients to the virtual clinic and hone telemedical tracking for those with worsening symptoms. AI might also optimize protocols for teleconsultation appointments at virtual clinics. It also contributes to the preparation of virtual clinical documents by pre-selecting details such as virtual clinic letters regarding patient and clinical findings and those that need immediate or urgent intervention. With anticipated demand and impact on healthcare consumers in view of short-notice cancellations of virtual clinical appointments, AI might drive the use of autonomous rerouting notifications to patients and also provide proposed rescheduled notices or handle rescheduling of consultations [17]. Telemedicine right now is at an inflection point, with the twin catalysts being the pressure of the pandemic and the long wait times for non-urgent in-office appointments for patients. There are other drivers too, not least the urging of patients not to visit physicians unless it is urgent or an emergency. With funding for virtual clinics being put in place in many systems, the use of data predictive models could enhance effectiveness, productivity, and, thus, resource utilization. We do not want to reduce clinical intensity per se, instead, we would like to see virtual clinics operate with a guaranteed optimum amount of increased throughput while maintaining, at the same time, high-quality patient care [4, 18].

1.2.1 Overview of Artificial Intelligence

The concept of AI is the focal point of several technological transformations, starting from personalized search engines to AI-driven autonomous cars, robotic arms, and drones. By using complex algorithms and data processing capabilities, otherwise known as models in AI terminology, AI is able to learn and aid in decision-making innately. Progressively evolved over the years, AI has transformed to suit the healthcare domain and has shown a multifold value in terms of enhancing patient care and optimizing clinical and management workflows. AI can be broadly categorized into two types based on knowledge or data performance: knowledge-based systems and data-driven AI [17]. The other way these systems are classified is based on the learning techniques—supervised, unsupervised, semi-supervised, reinforcement, and deep learning. The growing acceptance of AI stems from its ability to empower predictive analytics through longitudinal, multi-parametric data, which can subsequently be deployed in assisting clinical decision-making. In the pharmaceutical and biotechnology domains, AI has become invaluable in enhancing drug development and repurposing. Additionally, AI technologies have shown potential for significantly augmenting early diagnosis and treatment of chronic diseases using biomarkers and other imaging data. At an enterprise level, AI is paramount in streamlining operational workflows and improving patient care and management at hospitals [10]. Overall, AI serves as an effective tool in the non-invasive management of chronic diseases, particularly in the absence of diagnostic data, among other types of expert systems, as illustrated in Fig. 1.1.

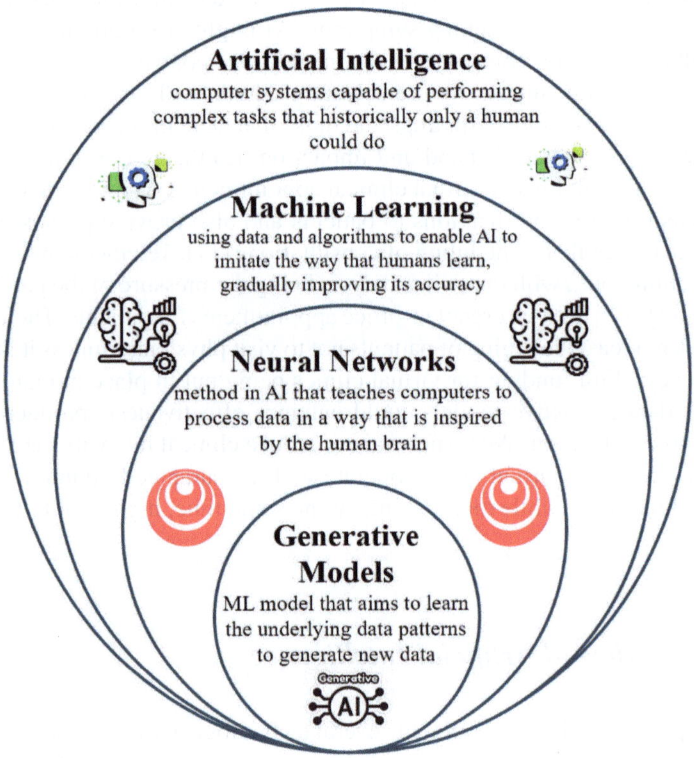

Fig. 1.1 Types of expert systems

1.2.2 Applications in Healthcare

Predictive modeling AI has been used in various sectors of healthcare to develop predictive models that can help in early detection as well as disease and case management. It has demonstrated that tools can predict diseases, future hospitalizations, as well as the localization of pain, with high accuracy. This has been a great deal of use because it can offer numerous benefits such as early prevention of disease, further disease management, individual care, and successful patient convenience in disease and pain diagnosis, treatment, and recovery rates. The development of administrative or clinical support tools can significantly improve the effective operation of healthcare settings [17]. Administrative support tools can produce remarkable outcomes, including the ability to predict the load of patient volumes, leading to effective resource management, the estimation of the need for healthcare services in particular disease encounters, with an increase or decrease depending on the environmental situation; and reduction of healthcare financial and staff waste. Clinical support tools, for instance, have been extensively developed in the field of imaging informatics. Recent years have witnessed rapid growth in the development and evaluation of AI tools for the rapid analysis of medical images through computer-assisted detection and diagnosis. Successful implementation of AI-computer-aided diagnosis (AI-CADx) has assisted radiologists and technicians in the cardiac field and has been one of the very few approved products by the newly established fourth central medical AI inspecting body and cleared as AI tools in the radiology and cardiology sections [19].

AI, particularly machine learning algorithms embedded within telemedicine systems, is constantly examined. AI-based personalized algorithms can learn over time through individuals' longitudinal data. The chatbot system has made a significant contribution to AI and telehealth users by building a robust platform that can synthesize the variety of chronic telemedicine use data to benefit both consumers and doctors. More specifically, it offers many automatic AI-based essential functionality tools that mine and transcribe our multiple mobile device-use chronic disease continuous data archives [11]. Limitless chatbot data mining could lead to large-scale insights in multimodal chronic disease user behavior; AI-driven summaries for both consumer and doctor, leveraging personalized data and system learning in addition to evidence-based models; data-driven predictions that combine population-based predictive models with personalized machine learning algorithms learned from prior consumer data to give a real-time update to an individual's telemedicine data; and extraction of embedded knowledge from massive individual health records [1]. Ethically, the flat case demonstrates that the risk allocation within telemedicine services can be significantly decreased in comparison with the stratified case. The study shows that AI diagnosis management tools should be further investigated to enhance chronic disease user trust in personalized telemedicine chatbot recommendations [20]. Given the imminent understanding of chronic disease user data, it is suggested that a trust line is recognized and shared with chatbot users for their ongoing chronic disease user preferences by implementing further AI management model developments and

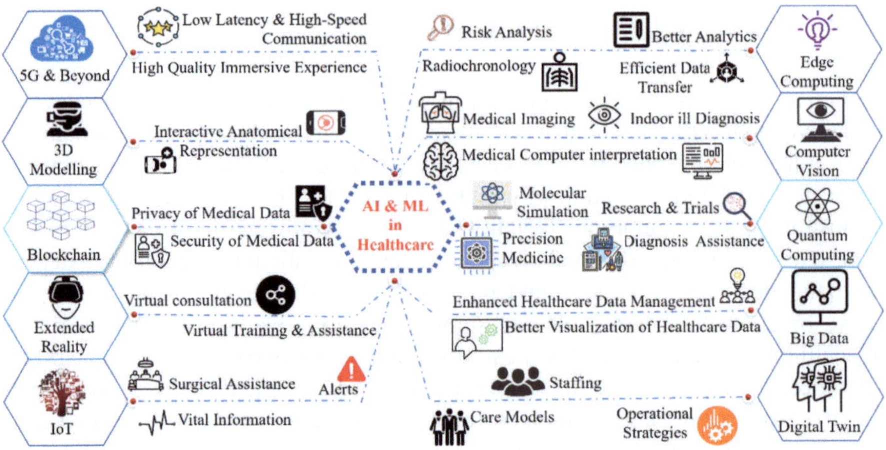

Fig. 1.2 Application of artificial intelligence (AI) and machine learning (ML) in healthcare

AI trust line explanations. Several applications of these technologies are presented in Fig. 1.2.

1.3 Telemedicine in Chronic Disease Management

Telemedicine refers to the delivery of health care services remotely by employing information and telecommunication technologies, including the internet, wireless and wireline communication, and satellite. It is concerned with an eclectic range of applications in general health care, specialty consultations, home health care, and population health management tools. However, the current review will focus on chronic care management from both developed and developing countries. Management of chronic diseases is defined in the morbidity chart, and a significant portion of health care spending is on chronic diseases, such as cardiovascular and cerebrovascular diseases, cancer, chronic respiratory diseases, diabetes, and oral diseases [1]. Chronic care management involves holding a patient's hand, i.e., frequent, targeted, and timely patient-physician interaction, where telemedicine plays a critical role. It aims at patient care at zero levels (such as walk-in clinics, doorstep services, and telemedicine).

Chronic disease management via home- and provider-mediated telemedicine has always been a focus not only for researchers but also for clinicians. The hallmark of chronic disease management through telemedicine is 'patient-physician interaction at a distance.' Newer areas, such as using sensors for monitoring, personalized monitoring, and handling extensive data, have also evolved. The actual surge in telemedicine practice began in the mid-twentieth century with the development of electrical and electronic recording and wireless technologies. The benefits of telemedicine for patients include reduced travel costs, less waiting time, fewer hospital visits, the ease of staying in contact with the physician during the recuperation period, the ability to work while monitoring their health, and the comfort and ease of care maintenance. Challenges in telemedicine include technical, legal, and regulatory issues, as well as patient interest and equity. Technical challenges include higher costs of equipment and low speed. The telemedicine regulatory area includes a need for higher medical liability, intensity, and privacy issues [10]. The upcoming problem relates to maintaining patient-physician relationships and privacy compliance issues. Additionally, patients in countries with a relatively lower telemedicine ratio and two or more prevalent chronic diseases are likely to avail themselves of telemedicine technology.

1.3.1 Understanding the Scope of Telemedicine

Telemedicine is the use of electronic information and telecommunications technologies to support long-distance clinical health care and patient and professional health-related education, public health, and health administration. It is imperative to understand the scope of telemedicine, as it is continuously evolving, especially since digital health technology and the patient-centered approach have considerably broadened the services under its wide umbrella. There are a variety of telehealth services provided, and the most common distinction is the type of interaction, such as, synchronous and asynchronous. Additional classifications are for services provided to particular locations, such as rural telehealth or in a hospital or clinic [5]. With increased internet speeds and smart technology, there is potential for telemedicine to take on a wider range of platforms associated with the technologies noted above. Services range from clinician-to-clinician, clinician-to-other staff, clinician-to-patient, patient-to-patient, and patient-to-other ones, like telepharmacy. Given that telemedicine is a broader term, we preferred to use telemedicine in the manuscript. Patients can be remotely located, close to medical facilities, home-based, or located in a hospital. Therefore, telemedicine cannot be just considered a method of providing remote care but also a service that has evolved to deal with and support a more mobile patient population, encompassing coordinated management of overall patient care, medical training, patient consultation via videoconferencing, transmission of still images, e-health including patient portal, remote monitoring of vital signs, continuing medical education and even communities affected by both natural and human-made disasters [6].

Providing more consultative services has evolved, with the recognition that a patient may not always be able to visit a healthcare provider, developing patient-centered services for which telemedicine is a front. Providing long-term care for the management of chronic conditions is one such valuable service. A growing body of evidence reflects its growing role; for example, a review reported nine of the ten included studies showed that telemedicine has been valuable for chronic disease management. Also, another review reported that telemedicine is a growing

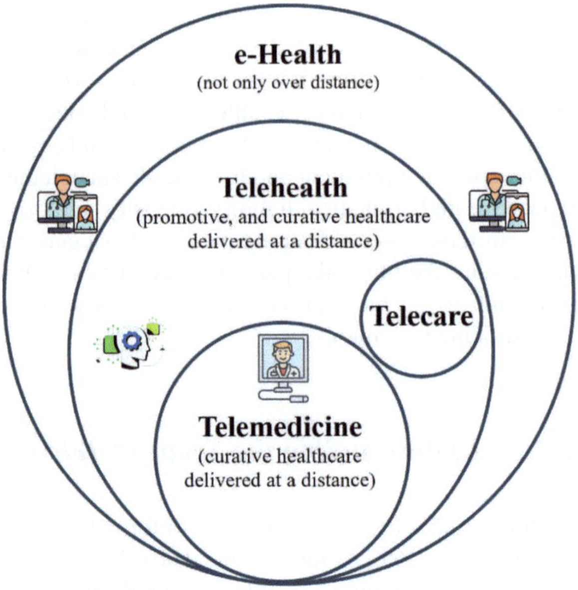

Fig. 1.3 Telemedicine and related concepts

part of the global e-health strategy aiming to manage multiple long-term medical conditions [17]. This aligns with a review showing the usability of telemedicine and benefits for patients and their quality of life were potential study outcomes for chronic diseases, such as chronic heart failure and diabetes. The term chronic disease management includes care for long-term curable and incurable conditions managed at the primary, secondary, and tertiary care settings. Many times, such cases are managed by general practitioners as well, and they also follow the concept of telemedicine in terms of regular consultation with the patients [15]. Providing regular meetings or routine visits reduces the severity of the disease in the long term and thus prevents any complications. It extends beyond telemedicine to providing telecare for individuals and telehealth, among others, as illustrated in Fig. 1.3.

1.3.2 Telemedicine Benefits and Challenges

Regarding chronic diseases, the benefits of telemedicine for acute care can also be expected at the level of chronic disease management. Even though robust evidence is not abundant, telemedicine has been shown to lead to better access to healthcare systems as well as more timely initiation or increase in treatment in conditions such as chronic obstructive pulmonary disease and heart failure. The greater convenience of remote consultation compared with face-to-face outpatient visits has potential benefits [21]. This greater convenience is postulated to be related not only to time saved by eliminating travel needs but also to flexibility in scheduling appointments due to the reduced time needed for travel so that patients would not need to take long periods off work or arrange for childcare. These, in turn, may lead to better adherence to the treatment plan and thus improve clinical outcomes, in addition to potential long-term benefits associated with adherence according to the disease being managed.

Limiting unnecessary travel and hospital attendance would also imply cost savings for patients, regardless of the distance from their healthcare service. It can be argued that telemedicine in chronic disease management strengthens the healthcare economy, provided that the effectiveness of any service's quality is maintained. Nevertheless, it is not without challenges; cost and the need to make additional provisions for poor patients or physically challenged patients may be one such challenge. The development or expansion of telemedicine services should be accompanied by a national, as well as international, discussion to ensure the freedom of its users from the burden of costs [18]. As with acute care, equity of access to telemedicine is still a concern since potential patients must have the ability to adapt to and use technology. Additionally, in some parts of the world, the provision of an internet connection can still be an issue. In line with acute medicine, privacy and confidentiality remain a paramount concern for telemedicine, especially with the far-reaching consequences of leaks. However, since the beneficiaries of such telemedicine are expected to be a more significant segment of society who are sicker, the potential impact of leaks can be perceived as harmful for a more significant number of people [22]. Given the early stages of this phenomenon and the factors above, many frameworks and implementation strategies of telemedicine in chronic disease management are being adopted to cope with it.

1.4 Integration of AI in Telemedicine for Chronic Disease Management

Telemedicine has become an integrated part of the healthcare industry, which can bridge the gap expanded by the urban–rural divide in healthcare services. However, managing chronic diseases remotely with a personalized approach is challenging, and the currently integrated solutions need to be backed up with more innovative technological interventions. The convergence of remote care management and AI focuses on clinical care in an individual-based approach that can be delivered from the comfort of patients' homes or remote care providers. AI and machine learning (ML) technologies progress and integrate with telemedicine to make remote chronic disease management smarter and to regularize future epidemics [23]. Several initiatives around the world are integrating AI with telemedicine and running small-scale pilot programs to manage chronic diseases remotely. The developed and integrated technology for AI and telemedicine is adaptive to learning with each patient's interaction. With the AI-powered patient-doctor service, a patient is provided with proactive care through the adaptive support delivered by healthcare professionals. AI technologies like natural language processing (NLP), machine learning (ML) algorithms, and cognitive learning enable decision-support systems and predictive models to be embedded for managing chronic conditions through telemedicine [24].

With this shift, AI-enabled telemedicine medications are becoming more common and are further influenced by the current pandemic to remotely treat acute and frequently chronic diseases. Trends based on AI and telemedicine are changing and becoming patient-centric. These trends are not just healthcare-centered but are indeed significant impacts of AI-enabled telemedicine within healthcare. It extends affordability and access to healthcare services. AI technology assists technology developers and stakeholders in introducing numerous initiatives and pilots in the healthcare market to enhance chronic disease management. It can skyrocket the business-generated revenue with positive population health. The proposed research paper conducted an in-depth review of the existing and emerging technologies and AI integration with telemedicine to manage chronic diseases [25]. Realistic application studies that integrated AI in clinical data and warm lines of communication demonstrated appreciable improvements in clinical outcomes and patient satisfaction in the context of managing chronic diseases. The proposed study of adaptive AI technology in chronic kidney disease and other routine chronic disease market scans will encourage healthcare researchers to develop adaptive telemedicine medications. This group of AI-integrated telemedicine can enhance therapeutic compliance technology, AI ethics, fairness, clinical decision support, and data integration. It also assists research enterprises in developing and integrating AI and telemedicine into existing healthcare methods [8]. However, telemedicine can be challenged differently depending on the application, and Fig. 1.4 illustrates sampled challenges of telemedicine in aviation situations.

1.4.1 Current Status and Trends

Multiple technological developments and numerous deployments in the telemedicine domain, specifically in chronic diseases, have indicated the feasibility of AI integration. At the academic level, there are a correspondingly large number of AI-related projects and pilot studies. These works range from the proliferation of digital health solutions to the validation of reliable and accurate AI models and devices for patient monitoring. Institutions, private companies, and governments have demonstrated a growing interest in liquidating pilot programs to validate AI research findings in the real world. Though telemedicine solutions have been around for the past two decades, the majority of pilot projects and national door-to-door campaigns have been launched over the last five years [6]. This reflects the growing attention paid by healthcare workers to AI solutions.

Recently, several research initiatives for innovative AI health have been launched and funded. These include programs involving the use of big data to synchronize research initiatives and strategies, pioneering cohort studies on the long-term health

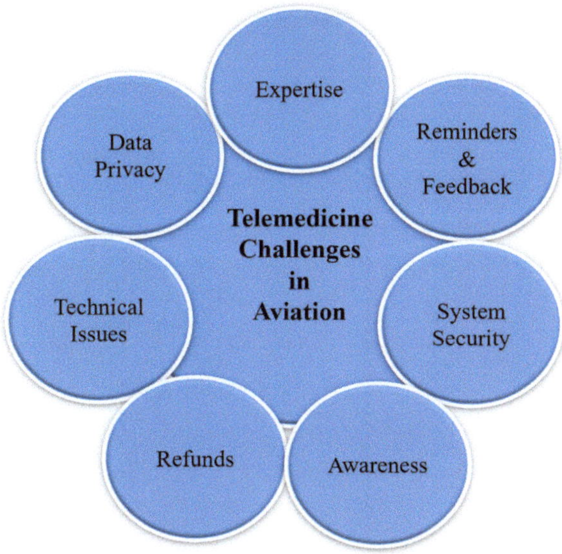

Fig. 1.4 Telemedicine challenges in the aviation arena

effects of certain substances, and consortia focused on the prevention of dementia. At the same time, market growth is forecast for global funding and investment in AI telemedicine, with a host of new technologies expected to emerge. Health service providers are confronted with high costs, mainly owing to the reimbursement of services developed using AI. The trend in investment and funding in entrepreneurship has strengthened interest within this domain, with a focus on project results and the start-up scene regarding telemedicine from an AI perspective [5]. Currently, regulatory efforts are steering this bullish trajectory. There have been statements regarding the use of advanced therapy medicinal products or other digital health technologies that leverage AI. Also, declarations on telemedicine and AI have been made at the organizational level, while there is a working group on digital health technologies and AI. Socio-nominal data surrounding patient acceptance of telemedicine and its job information is also quite varied, although none of the findings can yet be generalized to all chronic patients in society [21].

1.4.2 Key Technologies and Approaches

We first consider the significant stakeholders and components needed as building blocks towards a telemedicine-integrated AI solution. These requirements include remote monitoring, chat interfaces for patient-AI communication, and AI capabilities in analytics for patient data. The chat interface between AI and the patient can be made in the form of conversational agents or chatbots and has been exploited for managing specific conditions. In integrated predictive analytics, the algorithms

Fig. 1.5 Telemedicine use cases in aviation

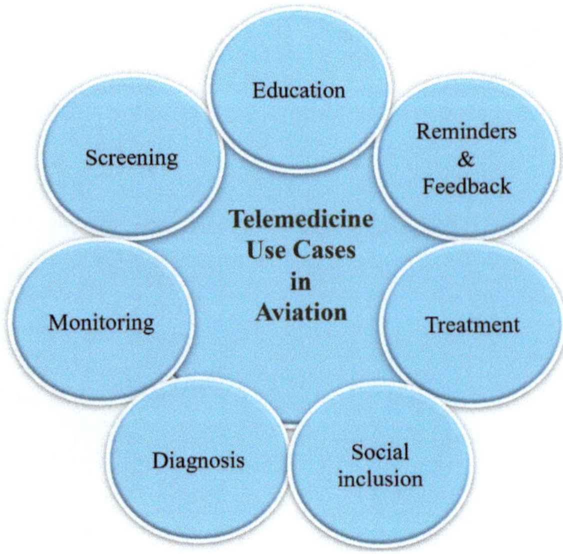

can make predictions of a patient's impending chronic condition flare and monitor the future progression of the condition as the patient follows a standard care plan. Telemedicine AI unlocks the potential of combining these solutions to place better analytics, diagnoses, and AI-driven care plans into the hands of the patient [25]. With remote devices resulting in large quantities of patient data, long-term AI monitoring and patient-generated health data analysis help AI model design and reduce clinical misdiagnoses, improving patient outcomes with some good use practices that can be attained, as depicted in Fig. 1.5.

Disease-specific telemedicine solutions use AI for personalized care plans, and we will discuss an example. These chatbots provide advice and support, keeping patients engaged in their healthcare journey. Like remote monitoring devices, these chatbots are adaptive and can direct the user to critical services if needed. There are also predictive analytics telemedicine platforms that provide algorithms to identify a range of possible chronic conditions for a patient who visits a GP. One approach that the system can take in response is to prescribe a set of future tests or investigations to narrow down the suspicion of certain diseases given additional patient data. Not surprisingly, this extensive platform provides for rapid interoperability with both existing platforms [4]. Finally, the platform is designed to reflect the speed of the art, and it incorporates the tools expected to become state-of-the-art through the lifespan of the platform and beyond. That is, the platform is a flexible diagnosis and prognosis system that can accommodate future advances. The system is currently being adapted to a standard patient-facing health chatbot. Continuous innovation is needed to keep up with the evolving healthcare needs. Research and development in the newly emergent Internet of Medical Things will likely pave the way toward the next generation of AI-driven telemedicine. We anticipate that the current AI will likely play a significant role in this next generation of healthcare, which focuses on preventing the patient's disease and coping with the social and environmental health determinants. Telemedicine AI will use either direct patient tracking signals or derive them from multiple non-health databases to reposition lifestyle apps and other consumer digital products that companies use to create overall health and well-being profiles of people for preventative care. Given the importance of TAI, it is crucial to capture these technologies as standard components of telemedicine [6]. This section, thus, provides a context for the reader's understanding of the current and potential future telemedicine paradigms available.

1.5 Case Studies and Success Stories

Within this section, sample case studies of telemedicine and its application are presented below.

1.5.1 Hypertension

The use of smartphone application alerts to remind hypertensive patients to take their antihypertensive medication has been a successful intervention in integrating AI and machine learning with telemedicine. These systems send reminders to the portable device, which can also measure blood pressure and send the readings to the healthcare professional's computer. Healthcare professionals can use AI-driven computer systems to generate recommendations and send them back to the computer on the patient's phone in order to optimize blood pressure management [18]. A combination of Pearson correlation and Spearman correlation on patients was used to characterize the correlation between their blood pressure measurements and, based on that, give them recommendations on how to manage their blood pressure. This resulted in an overall accuracy compared to the physician's recommendations. A proposed system where a feedforward backpropagation neural network system with two hidden layers can predict blood pressure values with a mean absolute error for systolic values and diastolic values, respectively. Moreover, the proposed system suggests medications for the patients and refers the result back to the healthcare provider. In their pilot study, different combinations resulted in the correct drug combination with a success rate for a group of patients having diabetes, hypertension, or both [25]. This high level of patient satisfaction measures a successful tool that can be adopted and used for blood pressure management.

1.5.2 Diabetes

In telemedical care for diabetic foot ulcers, a significant percentage of the study's participants, mainly patients and healthcare professionals, were either "satisfied with the foot image quality" or achieved a satisfaction score. These results show the significant opportunities of teledermatology and the confidence of the participants in the assessment. A proposed recommendation system for diabetic retinopathy type 2 patients suggested that, instead of telemedicine being conducted at specialized service centers with an ophthalmologic examination, a simple mail-out service should be provided directly to the consumers [19]. This mail-out tool could detect and help diabetic patients in the early stages of retinopathy development. The test group, consisting of a large number of patients, was informed about the service through ads. These patients determined the service for themselves and ordered any equipment they wanted. The analysis of the test group, including the path report and the images of the fundus, revealed that a percentage of the users sent images to their healthcare providers. When comparing the percentage of customer images for the three arms of the treatment, the most significant factor turned out to be the choice of individualism and liberty rather than the strategies and inducements that affected technology [6]. However, individualism was followed by the incentive of providing services and postal reminders. The final result is the implementation of a follow-up study to screen patients for treatment.

1.5.3 Clinic Pilot Programming

Several AI pilot deployments today began as early online services taken in clinic pilot programming for several obscure and metabolic indications that had significant unmet needs. These applications have evolved within the already established telemedicine company, and by incorporating a telemedicine service, the patients' satisfaction levels grew. For patients who have been happy with the AI shift and enjoyed that continuous method, this can be adjusted without having to travel to the clinic a couple of times per week. This is with an intuitive operator who did not require any substantial training or shift in the workflow of the clinic. As one would hope, non-pilot AI implementations will operate from a framework of present telemedicine platforms, which will permit modification and upload [5]. In several real-life implementations, specialized service sponsors paid to produce diagnostic AI or predictive technologies, while the telemedicine company has its proprietary AI tools that provide regular disease predictions and insights. Looking at the results of the implementation, some key findings were unanimous. First, the information shared in that session was excellent. Furthermore, there is a significant gap in automatic understanding. However, it's essential to offer a correct opinion on what the best action depends on. Consequently, AI algorithms using validated tools have the potential to facilitate frequent biomarker collections as well as patient engagement, as well as

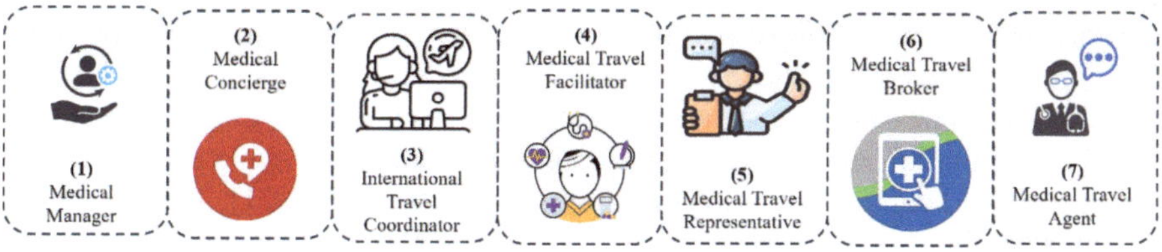

Fig. 1.6 Steps for performing telemedicine during a medical tour

to predict and prevent adverse clinical outcomes in larger groups of patients. Given the predictive costs needed to enhance the platform's results continually, the platform stakeholders highlight. Driven by patients and understanding the advantages, the payers have signed up for this service. The thought of maintaining an association with the business and assisting in delivering the most significant opportunities continues [25]. To attain smooth telemedicine operations, seven steps need to be followed, as summarized in Fig. 1.6, which starts with having a medical manager manage medical-related issues through a medical travel agent who assists online from the ground to those who are on the plane.

1.5.4 Impact on Patient Outcomes

Many studies have quantitatively and qualitatively evaluated the effectiveness of telemedicine with and without the use of AI in terms of its potential for enhancing outcomes, as well as in terms of benefits for adherence, costs, and experiences in using such interventions. We will emphasize the findings related to patient outcomes as the most important ones in this case. The implementation of AI in telemedicine can provide personalized care plans based on recent patient health or welfare status. In a qualitative analysis of case studies, we categorized one of the main impacts and effectiveness due to the use of telemedicine combined with AI as the enhancement of individual outcomes, such as lowering the number of inpatient days and improving quality of life, as well as enhancing adherence or changing patient ulfilmen. Results showed that the treatment effects differed significantly between the control and intervention groups [21]. In another study, AI-engine support using glucose prediction by continuous glucose monitoring reduced hyper- and hypoglycemic occurrences in children.

Furthermore, several case studies illustrated how telemedicine with AI components could enhance clinical outcomes, especially with the availability of real-time patient feedback. Studies elaborated that the use of AI in augmented reality effectively improves treatment outcomes, patient compliance, and the ulfilment of surgical goals and quality indicators, such as decreasing treatment errors and adverse events. Another study explored the potential of using a mobile application and AI support to monitor patient motion in the postoperative period, which successfully identified complications that occurred in patients with reductions in patient falls. Several studies in chronic patient case studies have also been able to establish a personalized rehabilitation plan. A report also mentioned a mobile application using AI, which is able to control difficult-to-register patients better [19, 21]. This means that the telemedicine tool has helped professionals identify clear recommendations for action and has sustained the program in the longer term.

1.6 Telemedicine Future Directions and Emerging Trends

AI in the coming years is expected to impact telemedicine in a number of ways. AI algorithms will likely continue to improve, which will lead to better chronic disease diagnoses and suggest treatments that reflect the best evidence and are personalized to the individual. Additionally, AI will contribute to the improved interoperability between telemedicine platforms and the various electronic medical record/electronic health record systems. Finally, the amount of big data will likely increase over time, which will lead to new treatment suggestions. Such data will allow for even better strategies to improve the active engagement of patients in their chronic diseases. Remote monitoring technologies and home-based digital health have been increasingly seen as part of the overall solution for chronic disease management [18]. AI is expected to underpin remote monitoring strategies with various wearables and sensors that collect gigabytes and terabytes of data on a per-patient basis while understanding what data are valuable and leading to improved patient outcomes. Telemedicine can improve care, but it is the technology that helps patients engage in their care that will have the most significant impact on disease management. However, there will likely be a significant need to improve AI algorithms to a high level of accuracy. Privacy and security may need to be worked into new software applications, and new regulatory frameworks will have to be established in order to integrate AI into everyday clinical care. As can be gleaned from these grand steps that show the potential for significant disruptive change, there is a continued need to engage all stakeholders [25]. Since some of the changes are here, stakeholders should begin to educate and adapt to the changes that are already occurring.

1.6.1 Advancements in AI and Telemedicine

Advancements in AI technologies in recent years have brought new possibilities for remote care in chronic disease management. Advances in ML have enabled clinically valuable applications in healthcare where large datasets are available, such as in diagnostics, prognostics, and predictive analytics. Innovations in natural language processing and computer vision now allow ML algorithms to nalyse and interpret unstructured patient data, such as clinician notes, medical images, and genomic analysis. AI-based chatbots and virtual health assistants now offer patients more conversational interaction, helping to monitor and manage their conditions. As telemedicine platforms integrating AI become more sophisticated and user-friendly, the options for using them to improve patient care will expand. AI has seen reasonable success in many clinical trials, and recent research seems to indicate that in a real-world healthcare setting, AI has promise [1, 11]. One study specifically found AI to be significantly better than physicians with experience ranging from one year to thirty years at diagnosing a disease from medical images. However, ML algorithms can be delicate, especially when they are deployed in new patient populations who have only subtle differences in their health signals. Additional challenges for integrating AI into telemedicine are the potential for algorithmic bias and the need to manage data privacy carefully. In the next decade, there are sure to be many advancements in the different areas of AI that directly apply to improving healthcare and healthcare technology, which could include breaking down complex disease processes into AI-interpretable models [17, 26]. Given the speed of advancements in these technologies, it is an excellent time to take advantage of them and leverage ML approaches to make telemedicine and remote medical monitoring a modern tool for patients and physicians in real time.

1.6.2 Potential Benefits and Risks

As we deploy AI, these two aspects are critical involving benefits and risks as we briefly present it.

Potential Benefits

The integration of AI in telemedicine has the potential to provide significant benefits to individuals with chronic conditions and the community at large. For individuals, the use of AI-supported telemedicine is expected to promote independence and improved patient outcomes by supporting condition self-management and promptly identifying problems or concerns before they escalate. For healthcare, it is expected that operating in a virtual model and personalizing care plans using day-to-day biomedical and personal data will make telemedicine more efficient and will allow individuals to be safely managed in the community, thereby minimizing hospital readmissions. By making care more accessible and efficient, the potential exists for a reduction in emergency department attendances, fewer withdrawals from community living to residential care, and a reduction in the high mental and emotional burden associated with chronic care provision [18, 21].

Potential Risks

The introduction of AI in telemedicine also comes with a number of potential risks. From an individual perspective, it will be essential to ensure that the use of AI does not compound information overload, a common source of frustration in chronic care, and a catalyst for withdrawal from contact with the healthcare system. Rigorous testing and ongoing evaluation of new AI systems will be needed to ensure that the analyses carried out are robust, reflect the needs and context of the user, and do not introduce bias. Such an approach will need to be sensitive to the ethical implications associated with integrating AI and other big data into healthcare delivery, such as the potential for breaches of privacy or security. From a healthcare perspective, potential risks include a false sense of safety that arises from the use of AI, the reliance on technology over clinical care, and a failure to respond to the individual and dynamically evolving needs of individuals. Moreover, concerns regarding algorithm error or bias point to a potential increase in legal disputes and negligence claims, in combination with the need to ensure that personal and biomedical data are kept secure. These potential risks reflect the need to approach the development and implementation of AI technology in a balanced and thoughtful manner and to stratify and share risk carefully [22, 25]. Importantly, this must be done in conjunction with adequate regulation and governance to ensure that the introduction of AI improves healthcare access and provision rather than compromising it.

1.7 Ethical and Legal Considerations

The implementation of AI in healthcare demands a few ethical and legal considerations, some as sampled in Fig. 1.7. Patients' consent to share medical and personal data and commit to decisions made by AI-driven algorithms is one of the critical points. Data privacy and security should also be ensured as telemedicine and digital health solutions are gaining increasing popularity. Similarly, the ethical use of patient data for AI purposes, also considering the legal implications, is a matter of ongoing debate. Who owns data in an AI-based application framework, and to whom is the consent given? Concerning AI software, specific regulatory frameworks governing medical AI and digital health software are under development in order to ensure the reliability, safety, security, and transparency of these systems [19, 24]. The general headings of these guidelines are: accountability and transparency, and data, information, and knowledge.

An existing limitation of AI is the selection of information available for the algorithms, which can lead to biases. Model biases are often related to individual biases, social and structural discrimination, and the method and accuracy of the available training data, reducing the generalization ability of the developed algorithms. For medical applications, ethical issues also include a possible disparity of treatment because of bias, raising concerns about treatment fairness and equity. The fairness

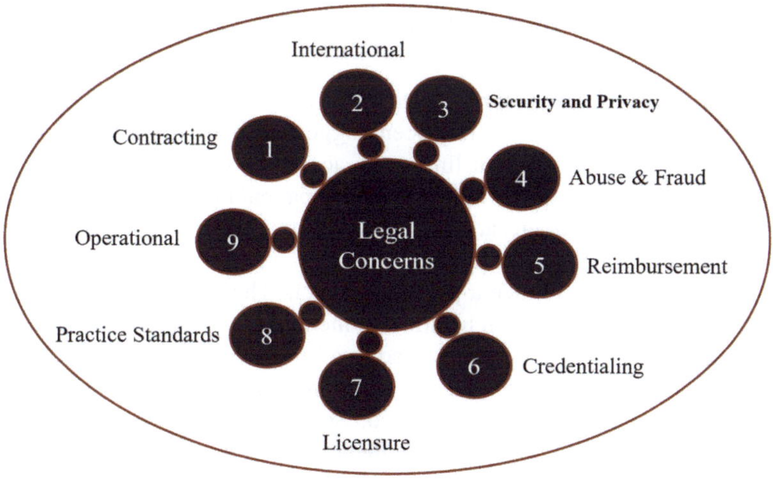

Fig. 1.7 Legal concerns about telemedicine

of AI algorithms has been explored in multiple AI applications. Ensuring transparency on how the AI systems work, therefore promoting the accountability of the purveyors of AI and the healthcare professionals who are entering this system, is another significant ethical consideration. AI solutions and workflows in healthcare should be explainable and justify the resulting decisions and recommendations, aiming for the meaningful inclusion of AI in a holistic clinical perspective rather than one of black-box algorithms [13, 24]. In addition, transparency has been indicated as a metric to evaluate the quality of the AI-assisted decision, as it reinforces the guarantee of the safety, security, and accuracy of the AI- and data-driven analysis. Regulatory frameworks for AI should envisage the establishment of a reliable and precise pipeline of accountability. Given the context of AI-assisted decision-making in healthcare, from the assessment of the eligibility of AI systems and their continual learning phase up to the authorization of a healthcare provider to use AI data and the regular post-market surveillance monitoring, the legal chain of accountability should ideally extend from the developer of the AI to the healthcare provider, in line with an independently supervised risk management system integrated into healthcare regulations [5, 6]. Resorting to digital twins of AI systems can also foster the standardization, reliability, and accountability of medical AI by prioritizing simulation over a full range of clinical relevance and low risk before AI transitions to in vivo and clinical studies.

1.7.1 Data Privacy and Security

Ensuring data privacy and security is the most crucial issue concerning the application of AI in telemedicine. It has been suggested that patient health data must be stored in a secure system that limits any access by unauthorized staff. Various strategies describe encryption to secure sensitive patient information at several steps, including input, usage, storage, and output. Additional security layers, such as secure communication channels, cybersecurity protocols, and virtual private networks, are also recommended to minimize a potential breach. The effectiveness of such tools is shown in many applications in healthcare, including radiology and pathology [8]. Failure to safeguard patient data may lead to a loss of patient trust in telemedicine, heavy fines, and legal liabilities for the institution involved, as has been observed when lawful regulations are not respected. Several case studies illustrate various forms of data breaches and their consequences. It is evident, therefore, that AI-driven telehealth tends to rely intensely on the compliance of both patients and healthcare providers with data protection regulations, regardless of local regulations and institutional guidelines. As the use of ICT introduces an increasingly dynamic environment of data sharing and teamwork within healthcare settings, healthcare providers, possibly in cooperation with the academic sector and industry, need to invest in developing cybersecurity cultures, heuristic strategies, and electronic risk minimization strategies to secure, but also restore, digital healthcare ecosystems [6, 24]. The discussion of data privacy and security does not aim to create a hostile view of AI or AI in telemedicine, this reflection, instead, aims to increase awareness of the importance of data protection, the correct use of AI in uncontrolled environments, and building compliance strategies.

1.7.2 Bias and Fairness in AI Algorithms

Bias in data that appears to be neutral is an ongoing crisis in healthcare data and for AI developers. In telemedicine, the training data may address acute and transient diseases, as opposed to diseases needing chronic disease management. Clearly, such data will produce an outcome for patients with chronic diseases that is different from when a chronic disease management system is used with the same dimensionality of the training data as the transient illness system. Such biases will be uncovered over time as the AI system is deployed, mainly when used in multi-racial, multi-ethnic, and rapidly growing communities, making these populations more likely to have unrecognized ancestral genetic, pharmacologic, and behavioral tendencies that the AI system may interpret as a disease state [3, 25]. Mitigating and eliminating such skews in AI outcomes will be difficult, as a wide range of demographic, metabolic, immunologic, and other patient and societal dynamics must be analyzed. Once understood, the next phase in AI telemedicine is to design and implement telemedicine systems informed by these software-centric results that adhere to societal humanist systems-based motivation relevant to social justice and health equity [21].

Ensuring fairness in AI is much more complex than simply “cleaning” the data of unintended biases, and no specific, universally accepted concept of fairness has emerged. Unsurprisingly, efforts to make AI systems fair inspire discussions around social, legal, and ethical considerations and may require years of debates to form the industry and societal criteria to govern these inconsistent datasets. The data inherent to these challenges stem from sociological and ideological backgrounds that need to be accounted for in solution management, but many are so rigid as to be impossible to modify. A necessary step to mitigate biases is to incorporate rigorous testing and validation into the development of AI algorithms by implementing guidelines and frameworks developed by the industry and policymakers. Periodic assessment of the outcomes and the impacts of AI-driven solutions must be prioritized, and the findings must be incorporated into the training processes through feedback loops [19, 24]. Given that the technology is still maturing, constant course corrections may be required while regulatory bodies continue to define the parameters for technical evaluation. The onus is on the developers and technology personnel to help in defining and developing these guidelines and to ensure these guidelines align with the obligations of the medical community to serve their patients first, avoid harm, and treat all patients with equity. Simply meeting technical righteousness does not fulfill the ethical import of the popular culture terms of either medicine or society. This biodiversity may have naturally evolved with Homo sapiens representing gender, ethnicity, social status, and sociodemographics, but the 21st-century intersectionality expects the same AI for all, “uniformity wrapped in equity as social justice" [18]. Sustainable technologies that support telemedicine are presented in Fig. 1.8. These tools provide alerts and metadata of the patient instantaneously.

1.8 Recommendations

Healthcare professionals need to stay abreast of developing guidelines for best practices in using AI in telemedicine while ensuring compliance.

1.9 Summary of Key Findings

Throughout the previous sections, it has become clear that AI holds immense promise for improving chronic disease management and many other areas of healthcare. Notably, the growing incorporation of telemedicine offers a range of potential contributions that AI can make to enhance care and lower costs. These include improved patient engagement, more effective monitoring of patient’s health status and adherence to treatment plans, better patient-centered care, and the ability to expand the reach of specialists in urban centers. Throughout these applications, AI can customize

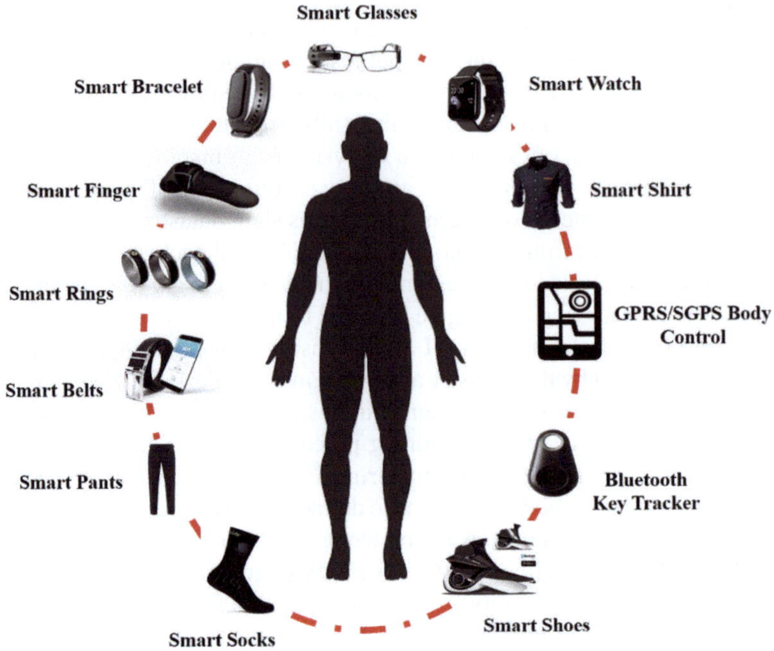

Fig. 1.8 Sample innovative enabling technologies of telemedicine

patients' self-management and can also learn clinical data from disparate sources. Despite these potential benefits, significant challenges to the use of AI in chronic disease management continue to exist, including data privacy and patient consent, context sensitivity, fear of replacing healthcare providers, care silos, and patient adoption issues [13]. Many of these challenges could be addressed through thoughtful policy-making and research, as well as innovative technology features that will not decrease the ethical standards and patient-provider relations in healthcare practices.

This literature review has shed light on several key themes. Throughout the chapter, we explored the potential benefits of using AI in chronic disease management, especially when in combination with telemedicine; for example, Fig. 1.9 presents an existing telemedical application scenario using the medical Internet of things. Given the interest in personalized medicine and precision health, AI is likely to play an increasingly important role in research in these domains. We described a range of AI methods, tools, and devices and gave examples of how they are being used in research during the pandemic. Looking ahead, an area for further research is the combination of AI and telemedicine technology, for which the data is only beginning to accumulate [6]. We also explored some of the current challenges constraining AI's potential to impact chronic disease management. In this respect, more research is needed on how well data gleaned in specific populations during telemedicine consultations generalizes to larger patient populations and other healthcare practices. Nonetheless, by focusing on these and many other related topics, telemedicine

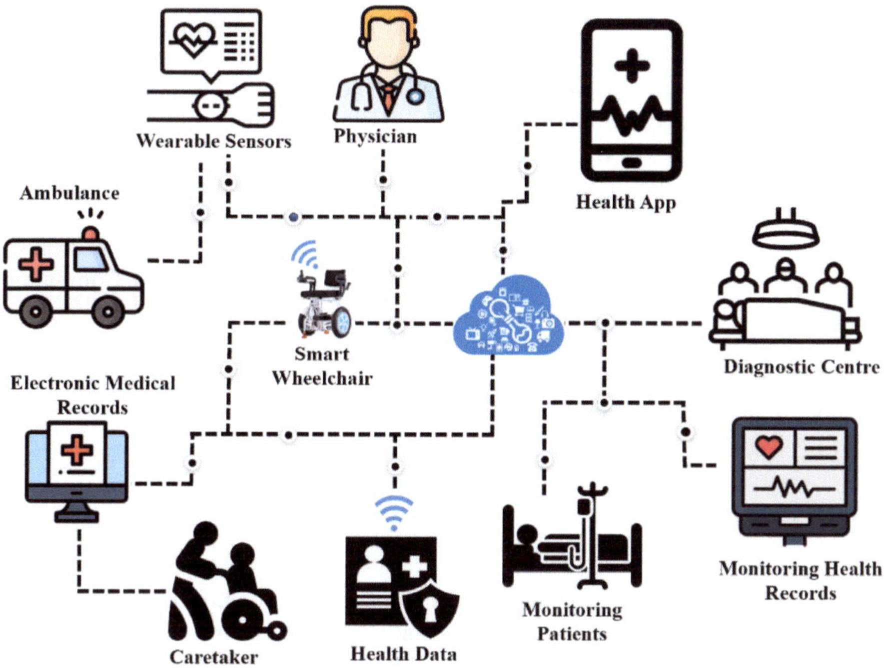

Fig. 1.9 A telemedical application scenario using the medical Internet of things

is on course to change the way we think about and conduct research in the future. The future will involve a multidisciplinary approach involving engineers, data scientists, clinicians, patients, and many other stakeholders interested in improving data collection and analysis during telemedicine consultations [19].

1.9.1 Implications for Healthcare Practice

The integration of AI in telemedicine represents a transformative opportunity for healthcare. Many models for improved chronic disease management are based upon reactive care models; however, new technologies can create personalized models for care that will see many more patients achieving optimal outcomes. These technologies will improve the way that care is provided, and as our clinical workflow can adjust to that, outcomes are likely to be better. A new generation of healthcare professionals who grew up in a digital-native world will take these innovations in their stride. However, aspects of training, interdisciplinary work, and modes of collaboration between individuals delivering care are due for substantial overhauls [5]. Fully embracing these new technological models will be necessary to maximize benefits from them while minimizing disruptions to ongoing patient care. As values of informed consent, patient data privacy, and justice in healthcare remain central to policies to protect patients, similar guidance and regulations accompanying the development and use of AI in healthcare must be developed. The progressive technology adoption perspective described here also implies and entails that ongoing societal disparities will exist for some time. Communities with less access to and understanding of technology may be left behind in healthcare improvements. Proactive models for educating and training individuals in the use of digital systems, reading AI-driven clinical information, and so forth will likely make use of existing community assets and, in turn, strengthen communities from within [27]. Community empowerment through such educational and training programs can form a catalyst for the adoption of evolving telehealth systems. Providing these systems will not only improve healthcare but will also shore up patient communities to improve their response and preparedness in the case of any future pandemic.

1.10 Conclusion

Our findings from the reviewed literature demonstrate the potential of AI in the effective management of chronic diseases, especially when integrated with telemedicine. Several AI and machine learning algorithms have been introduced for chronic disease management. Team-based discussions should center on potential unintended consequences of AI employment and develop strategies to evaluate whether these occur. Plans should also include a posteriori data analysis to investigate the impact different clinical, social, or demographic factors may have on model accuracy and potential bias in the training data and for defining ethical guidelines needed for the use of AI in clinical practice to ensure that AI-driven predictions are not used to perpetuate inequities and disparities in the way that health services are delivered. Evidence from high-quality studies of telemedicine in general and other innovative remote technologies suggests that these digital revolutions have the potential to fill the gap in chronic disease management. This also implies that integrating telemedicine-based AI technologies can enhance chronic disease care delivery and management. This integration can open a new pathway in healthcare delivery where human intelligence is synergized with AI, bringing unprecedented precision and predictive power that were previously impossible to attain. Drawing on the extensive literature, we recommend several courses for the way forward. First, AI-enabled telemedicine should be addressed from an all-inclusive approach. Experience has shown that the most significant impact we can have in healthcare digitalization is a broader, inclusive alternative approach. From our review, we found that ensuring efficient AI in chronic disease management is requisite for early and sustained collaboration among all relevant stakeholders, including patients and their care partners. Second, a series of research must be initiated that would specifically provide comprehensive information about the effectiveness and precision of AI-enabled telemedicine in diverse populations. Third, ethical issues and the means of ensuring compliance will be crucial for the responsible deployment of AI in telemedicine.

References

1. Manju, R., Harinee, P., Gangolli, S. S., & Bhuvana, N. (2023). Evolution of Computational Intelligence in Modern Medicine for Health Care Informatics. In *EAI/Springer Innovations in Communication and Computing: Vol. Part F282*. https://doi.org/10.1007/978-3-031-27700-9_24
2. Shafik, W. (2024a). Artificial Intelligence and the Medical Tourism. In *Examining Tourist Behaviors and Community Involvement in Destination Rejuvenation* (pp. 207–233). IGI Global. https://doi.org/10.4018/979-8-3693-6819-0.ch016
3. Tsoi, K., Yiu, K., Lee, H., Cheng, H. M., Wang, T. D., Tay, J. C., Teo, B. W., Turana, Y., Soenarta, A. A., Sogunuru, G. P., Siddique, S., Chia, Y. C., Shin, J., Chen, C. H., Wang, J. G., & Kario, K. (2021). Applications of artificial intelligence for hypertension management. In *Journal of Clinical Hypertension* (Vol. 23, Issue 3). https://doi.org/10.1111/jch.14180
4. Shafik, W. (2024b). Artificial Intelligence-Enabled Internet of Medical Things (AIoMT) in Modern Healthcare Practices. In *Clinical Practice and Unmet Challenges in AI-Enhanced Healthcare Systems* (pp. 42–69). IGI Global. https://doi.org/10.4018/979-8-3693-2703-6.ch003
5. Ma Pilar, V. T., & Mateu Hernández, C. A. (2021). Bridging the technology gap of telemedicine. *Atencion Primaria Practica, 3*. https://doi.org/10.1016/j.appr.2021.100117
6. Park, Y., Lee, C., & Jung, J. Y. (2022). Digital Healthcare for Airway Diseases from Personal Environmental Exposure. *Yonsei Medical Journal, 63*. https://doi.org/10.3349/YMJ.2022.63.S1
7. Shafik, W. (2024c). Connected healthcare—the impact of Internet of Things on medical services: Merits, limitations, future insights, case studies, and open research questions. In *Artificial Intelligence and Internet of Things based Augmented Trends for Data Driven Systems* (pp. 181–217). CRC Press. https://doi.org/10.1201/9781003497318-10
8. Szelągowski, M., Berniak-Woźny, J., & Lipiński, C. (2021). Bpm support for patient-centred clinical pathways in chronic diseases. *Sensors, 21*(21). https://doi.org/10.3390/s21217383
9. Shafik, W. (2024d). Deep Learning Impacts in the Field of Artificial Intelligence. In *Deep Learning Concepts in Operations Research*. (pp. 9–26). Auerbach Publications. https://doi.org/10.1201/9781003433309-2
10. Li, T., Zhao, Y., Sun, L., & Feng, C. (2021). Artificial intelligence medical platform and clinical care of chronic obstructive pulmonary disease based on microprocessor. *Microprocessors and Microsystems, 82*. https://doi.org/10.1016/j.micpro.2021.103879
11. Dibie, A., & Ingremeau, D. (2021). Chronic Heart Failure and Telemedecine, remote monitoring: Interest and perspectives. In *Annales de Cardiologie et d'Angeiologie* (Vol. 70, Issue 5). https://doi.org/10.1016/j.ancard.2021.08.007
12. Shafik, W. (2024e). Digital healthcare systems in a federated learning perspective. In *Federated Learning for Digital Healthcare Systems* (pp. 1–35). Academic Press. https://doi.org/10.1016/B978-0-443-13897-3.00001-1
13. Niko Azhari Hidayat, Abed Nego Okthara Sebayang, Amila Shofia, & Shofa Aulia Aldharma. (2022). DEVELOPMENT OF TELEMEDICINE APPLICATION BASED ON ARTIFICIAL INTELLIGENCE IN THE HANDLING OF DIABET FOOT. *Jurnal Layanan Masyarakat (Journal of Public Services), 6*(1). https://doi.org/10.20473/jlm.v6i1.2022.205-211
14. Shafik, W. (2024f). The Future of Healthcare: AIoMT—Redefining Healthcare with Advanced Artificial Intelligence and Machine Learning Techniques. *Artificial Intelligence and Machine Learning in Drug Design and Development*, (pp. 605–634). Scrivener Publishing LLC. https://doi.org/10.1002/9781394234196.ch19
15. Alcoceba-Herrero, I., Coco-Martín, M. B., Leal-Vega, L., Martín-Gutiérrez, A., Peña-de Diego, L., Dueñas-Gutiérrez, C., de Castro-Rodríguez, F., Royuela-Ruiz, P., & Arenillas-Lara, J. F. (2023). Randomized Controlled Trial Evaluating the Benefit of a Novel Clinical Decision Support System for the Management of COVID-19 Patients in Home Quarantine: A Study Protocol. *International Journal of Environmental Research and Public Health, 20*(3). https://doi.org/10.3390/ijerph20032300

16. The U.S. Food and Drug Administration. (2020). What is digital health? In *Digital Health Center of Excellence*. https://www.fda.gov/medical-devices/digital-health-center-excellence/what-digital-health
17. Verma, N., Vojjala, N., Mishra, S., Valsan, A., Kaur, R., Kaur, T., De, A., Premkumar, M., Taneja, S., Duseja, A., Singh, M., & Singh, V. (2023). Machine learning can guide suitability of consultation and patient referral through telemedicine for hepatobiliary diseases. *Journal of Gastroenterology and Hepatology (Australia)*, *38*(6). https://doi.org/10.1111/jgh.16194
18. Geetanjali, Malviya, R., Awasthi, R., Sharma, P. K., Kala, N., Kumar, V., & Yadav, S. K. (2022). Applications of artificial intelligence, blockchain, and internet-of-things in management of chronic disease. In *Cognitive Intelligence and Big Data in Healthcare*. https://doi.org/10.1002/9781119771982.ch13
19. Brandão-De-Resende, C., Alcântara, L. D. A. R. De, Vasconcelos-Santos, D. V., & Diniz-Filho, A. (2023). Glaucoma and Telemedicine. In *Journal of Glaucoma* (Vol. 32, Issue 5). https://doi.org/10.1097/IJG.0000000000002200
20. Seneviratne, M., & Peiris, D. (2019). Digital health in low- and middle-income countries. In *Revolutionizing Tropical Medicine: Point-of-Care Tests, New Imaging Technologies and Digital Health*. https://doi.org/10.1002/9781119282686.ch32
21. Steiropoulos, P., Bakakos, P., Hatziagorou, E., Katsaounou, P., Loukides, S., Papaioannou, A., Porpodis, K., Samaras, K., Tzouvelekis, A., Kalafatakis, K., & Kostikas, K. (2021). The present and future of inhalation therapy for the management of obstructive airway diseases: Emphasis on pressurized metered-dose inhalers. *Pneumon*. https://doi.org/10.18332/pne/144614
22. Andrès, E., Meyer, L., Zulfiqar, A. A., Hajjam, M., Talha, S., Bahougne, T., Ervé, S., Hajjam, J., Doucet, J., Jeandidier, N., & El Hassani, A. H. (2019). Telemonitoring in diabetes: evolution of concepts and technologies, with a focus on results of the more recent studies. *Journal of Medicine and Life*, *2019*(3). https://doi.org/10.25122/jml-2019-0006
23. Jedamzik, S. (2019a). Digital health and nursing: The future is now. *Pneumologe*, *16*(6). https://doi.org/10.1007/s10405-019-00279-4
24. Jedamzik, S. (2019b). Digital health and nursing: The future is now. In *Unfallchirurg* (Vol. 122, Issue 9). https://doi.org/10.1007/s00113-019-0672-2
25. Bhatt, P., Liu, J., Gong, Y., Wang, J., & Guo, Y. (2022). Emerging Artificial Intelligence-Empowered mHealth: Scoping Review. In *JMIR mHealth and uHealth* (Vol. 10, Issue 6). https://doi.org/10.2196/35053
26. Kong, L. Y., Ramirez-Garcialuna, J. L., Fraser, R. D. J., & Wang, S. C. (2021). A 57-year-old man with type 1 diabetes mellitus and a chronic foot ulcer successfully managed with a remote patient-facing wound care smartphone application. *American Journal of Case Reports*, *22*(1). https://doi.org/10.12659/AJCR.933879
27. Srivastava, M., Siddiqui, A. T., & Srivastava, V. (2023). Application of Artificial Intelligence of Medical Things in Remote Healthcare Delivery. In *Handbook of Security and Privacy of AI-Enabled Healthcare Systems and Internet of Medical Things*. https://doi.org/10.1201/9781003370321-7
28. Shafik, W. (2024g). Wearable medical electronics in artificial intelligence of medical things. *Handbook of security and privacy of ai-enabled healthcare systems and internet of medical things*, 21-40. https://doi.org/10.1201/9781003370321-2
29. Shafik, W. (2025a). Revolutionizing Skin Cancer Diagnosis with Artificial Intelligence: Insights into Machine Learning Techniques. In *Impact of Digital Solutions for Improved Healthcare Delivery* (pp. 167–194). IGI Global. https://doi.org/10.4018/979-8-3693-5237-3.ch006
30. Shafik, W. (2025b). Smart biomedical devices for smart healthcare. In *Machine Learning Models and Architectures for Biomedical Signal Processing* (pp. 421–448). Academic Press. https://doi.org/10.1016/B978-0-443-22158-3.00017-X

Chapter 2
The Integration of AI and Telemedicine in Chronic Illness Care: Current Trends and Future Insights

P. R. Anisha, Kaniti Anjali, and V. Sridhar Reddy

Abstract With the transformative role of Artificial Intelligence in chronic disease management, telemedicine and AI-driven tools enable early diagnosis, personalized care, and continuous monitoring. In the Healthcare 5.0 framework, sustainable telehealth systems enhanced by AI facilitate real-time data analysis and predictive insights, advancing clinical outcomes and optimizing resource use. The paper presents a thorough examination and analysis of AI-assisted telemedicine interventions concerning chronic disease management through case studies: Propeller Health for asthma, MySugr for diabetes, PathAI in pathology insights, DeepMind Health for predictive health diagnostics, and BenevolentAI for oncology, emphasizing their approaches, efficiencies, and outcomes. This study comparatively analyzes the above case studies against notable advances they made in patient care, including enhanced diagnostic accuracy and personalized treatment plans. Key contributions include a comprehensive evaluation of supervised and unsupervised machine-learning models to predict the disease, evaluations of telehealth solutions for AI-enabled remote patient monitoring, and examples of how machine learning algorithms have enhanced diagnostic accuracy and individualized treatment plans. Our systematic approach from the methodology on study selection and evaluation itself assures the reliability and relevance of the findings. Findings suggest that these intelligent technologies hold significant promise in chronic disease management but run into real issues of data privacy, scalability of models, and adaptability in varied healthcare settings. Through discussions on current outcomes, implications, limitations, and future directions, aims to provide a strategic framework for incorporating AI in telemedicine, encouraging the development of flexible, secure, and ethically driven AI models for healthcare deployment. This research ultimately contributes to the scientific discipline, strengthening that vision of adaptively advanced AI models that will enhance access to high-quality healthcare and raise efficiency, ultimately paving the way for sustainable data-driven healthcare interventions. By exploring the potential of AI to

P. R. Anisha · K. Anjali (✉)
Stanley College of Engineering and Technology for Women, Hyderabad, India
e-mail: kanitianjali2005@gmail.com

V. S. Reddy
Vignana Bharathi Institute of Technology, Hyderabad, India

© The Author(s), under exclusive license to Springer Nature Switzerland AG 2026
K. Dalip et al. (eds.), *Healthcare 5.0*, https://doi.org/10.1007/978-3-032-09582-4_2

significantly improve patient outcomes and reduce healthcare costs, this paper advocates for innovations that address accessibility challenges in chronic disease management. Ultimately, the findings point out the possibility for AI and telemedicine to have a greater role in access to individualized care, thus helping close existing care gaps while optimizing chronic care management solutions within a patient-centered context.

Keywords AI in healthcare · Chronic disease · Telemedicine · Healthcare 5.0 · Sustainable telehealth · Remote patient monitoring

2.1 Introduction

Noncommunicable Diseases (NCDs), such as cancer, diabetes, arthritis, chronic respiratory diseases, and cardiovascular diseases are the leading and major cause of permanent disability and even premature deaths [1]. As per the World Health Statistics Report 2020 released by World Health Organization (WHO), premature deaths from NCDs (between the ages of 30 and 70) occur in 41 million people annually, which is about 74% of all deaths globally [2]. If the current trend follows, by 2030, this figure could reach 84%. In the United States, six out of ten people have at least one chronic disease, and about four out of the ten have two or more, which highlights the fact that NCDs are a serious global issue. Major of the NCDs can be prevented or controlled to some extent but not always completely treatable. This highlights the need to address the importance of inculcating some important chronic diseases prevention and management strategies.

Most of these premature deaths are preventable, if health systems are able to respond more effectively to people's healthcare needs. To control the effect of chronic diseases, the WHO established the Global Action Plan 2013–2020, which addresses nine target principles. A tangible way to contribute towards achieving these targets is through integration of emerging advance technology with healthcare, transforming into a smart healthcare solution [3].

"The magic of Artificial Intelligence lies not in what it can do today, but in how it will transform the healthcare of tomorrow."- Paul Sonnier. With rapid advancements in AI, modern intelligent systems have always accompanied humans in the field of healthcare by assisting in tasks that otherwise would either consume more time or assistance of a medical expert or demands precision [4, 5]. The Healthcare 5.0 integrates cutting-edge advanced technologies with human intelligence and skills to revolutionise healthcare delivery, emphasise automation, precision medicine, seamless connectivity between the doctors and the patients, thus, enhancing overall quality of healthcare delivery. With the help of Sustainable Telehealth, an economically and environmentally sustainable remote access and delivery of healthcare services could be achieved.

For example, AI systems can help in regular monitoring of vital health data of patients under proper expertise [6]. By analyzing vast amounts of patient data, AI algorithms can identify patterns and trends that human clinicians might miss. This predictive capability allows for earlier interventions and more precise monitoring [7, 8]. The AI system can be integrated in wearable monitoring devices such as glucose monitors and smart watches, allowing remote health care access to the people [9]. For Example, WioMT (Wearable Internet of Medical Things) are the devices that interconnect wearable sensors to monitor human health, behavior and wellness. These devices such as WioMT, are developed by the researchers for monitoring and prevention of NCDs. Its purpose in healthcare systems is to provide continuous health monitoring, collecting real-time data, detection of early symptoms of chronic conditions such as diabetes and hypertension [10].

Consequently, several AI techniques can be used for inspection to retrieve patterns from data [11, 12]. AI can assist in creating personalized treatment plans, based on the health data collected from monitoring the patient such as the eating habits, lifestyle, genetics and medical history. This will ensure that every patient receives an effective treatment, reducing the risks of increased complications and improving overall health. A notable example is Phable, an Indian startup, provides a telehealth platform for chronic disease management. This platform's smartphone application allows patient to connect with doctors for cardiac problems, cancer treatment or diabetes and provides access to Internet of Things (IoT) medical devices for vitals monitoring. This application shows a good combination of technology and medical science that combines data extracted from the tech devices with the expertise opinion of real-time professional expert doctors. Moreover, CareClinic, a Canada-based startup, introduced a self-care health management platform for people suffering from chronic conditions [13]. The application, powered by artificial intelligence, empowers users to track their symptoms, mood, pain level, and nutrition levels. This enables patients to take a control of their own health and make an informed decisions.

Physicians spend only about 25% of their time on direct care of patients, while about 31% of the time is spent on the documentations and paper works, consuming most of the time in the activities that do not help in direct care of the patients, thus increasing the wait time of patients who pay regular visits to clinics due to chronic medical conditions [14]. Healthcare 5.0 with a Sustainable Telehealth platforms provides an innovative approach to the Chronic Disease Management, by enabling continuous remote monitoring, consistent access to care, reducing the need for regular hospital visits, lowering the overall healthcare costs and maximizing the quality of outcomes for patients dealing with conditions like diabetes, heart diseases, arthritis or hypertension through virtual AI-driven platforms. The ultimate goal is to redefine and improve chronic disease management through technologies such as cloud computing, data-driven applications, artificial intelligence (AI), and the Internet of Things (IoT) by extending digital capabilities in three main areas: personalized healthcare facility, patient-centered solutions, and seamless access to a variety of telemedicine services and health knowledge.

The significant contributions of this research are:

- An in-depth comparative analysis of five notable AI-supported telemedicine interventions persistently in the chronic disease management demonstrates considerable improvements in patient care, diagnostic accuracy, and personalized care planning.
- By utilizing real-world applications such as Propeller Health, MySugr, PathAI, DeepMind Health, and BenevolentAI, the paper reviews the technological advancements made in the various fields of chronic medicine-from caring for asthma to caring for cancer.
- The research uses a structured and transparent methodology for choosing, evaluating, and analyzing the studies based on certain criteria providing valuable insights into the integration of AI and telemedicine in chronic disease management.
- In-depth discussion of the outcomes, implications arising from the state of the research, limitations, and future directions will guide stakeholders in integrating AI into telemedicine for chronic care to enhance patient outcomes through early intervention, continuous monitoring, and wellbeing personalized strategies.
- The paper takes an optimistic view on AI in health, particularly in the area of chronic disease management, signaling the future directions that will fill the present gaps and explore even newer horizons with telemedicine and AI interleaved.

2.2 Literature Survey

Chronic disease management has witnessed a remarkable change over the years, highlighted by both gradual transformations and radical breakthroughs in methodology. In this literature survey, studies on technologies in chronic disease management along with other related works and surveys are investigated and their approaches, contributions as well as limitations are demarcated.

Initial Hybrid Techniques for Chronic Disease Management

Hybrid models were implemented, employing a vast selection of machine learning and classical statistical techniques in tackling real-time blood glucose monitoring in type 1 diabetes patients [15]. Bridging the gap between classical statistical models and artificial intelligence, their work achieved an advanced time point for insulin administration. However, variance in patients' metabolism constituted obstacles on their path, inspiring future tasks toward adaptive personalized models.

Advanced Models and Intensive Applications

Studies were focused on the use of advanced machine learning techniques and AI solutions for chronic diseases management. This period is viewed as a significant leap toward the application of AI in real-world scenarios, such as wearable technologies, medical imaging, and electronic health records- aimed at improving diagnostic accuracy, early detection, and personalized treatment strategies. Though each study targeted different domains, their underpinning generally pointed towards the prospect of AI and, at the same time, outlined the flaws that need addressing in order to further improve the tools going forward.

Supervised machine learning models were employed for chronic heart diseases with data gathered from wearable devices alongside algorithms such as Random Forest and XGBoost [16]. This was a step toward integrating AI into wearable technologies to improve diagnostic accuracy concerning cardiac health. Demographic diversity was noted as a limitation, as results were highly variable between demographic groups, suggesting that the datasets need to be more expansive and inclusive.

CNNs (Convolutional Neural Networks) were used in conjunction with clinical diagnostic data on images. Wang et al. [17] used AI-assisted imaging processing techniques in diabetic retinopathy. Its accuracy for early-stage detection exceeded the traditional imaging techniques, but the most limiting were the quality and consistency of the inputs.

The analysis of the medical records of patients for the early identification of chronic obstructive pulmonary disease was conducted using deep learning models [2]. A significant sensitivity and specificity was achieved much early in the COPD complaint, which added to its credibility. However, limitations included concerns about data privacy and varying data quality due to inconsistent record-keeping among different healthcare facilities.

A method was particularly developed for chronic kidney disease, where reinforcement learning was used to schedule dialysis treatment integrating patient's history to optimize time and reduce readmission [18]. The outcome of this study was a promising facility that does not lose sight of the challenges faced, which include the necessity for highly personalized datasets and high computational power, inhibit any real-world implementation in standard hospital settings.

Exploring Unsupervised Learning and Its Potential

Studies were conducted on several machine learning models for disease risk prediction using unsupervised machine learning models. Lu and Uddin [19] used unsupervised machine learning models within seven algorithms in 15 datasets (heart failure, diabetes, breast cancer) for chronic disease stages and risk predictions. Among them, DBSCAN stood out with good Homogeneity, Completeness, and V-measure; however, no one model performed best on all datasets. The findings demonstrate the potential in unsupervised machine learning when labeled data is limited, as in the domain of undiagnosed diseases. A limitation here is that well-documented disease records would be required to really fine-tune model evaluation and selection, as outlined in future works.

Over the years, AI integrated telemedicine has seen its evolution from classical to hybrid approaches, then to specialized models and finally to real-world integrations. Key developments include the use of wearable devices and electronic health records, unsupervised and reinforcement learning, and early diagnosis and personalized care. However, limitations such as data privacy issues, differences in demographics, and disparities in data quality continue to exist; that our research aims to address comprehensively.

2.3 Case Presentation

2.3.1 Introduction to the Study Groups

The five case studies deal with chronic diseases such as asthma, COPD, diabetes, cancer, chronic kidney disease (CKD), and rheumatoid arthritis (RA). Each group consists of patients diagnosed with these different chronic diseases, with different demographic variables such as age, sex, and comorbidities.

Propeller Health (Asthma & COPD): Primarily develop asthma and COPD, usually middle-aged or older, frequently with smoking-related respiratory diseases. A number of patients manifest comorbidities, notably cardiovascular diseases.

MySugr (Diabetes Management): It mainly manages Type 1 and Type 2 diabetes, including patients of different ages, but mainly middle-aged adults. The population suffers from comorbidities, such as obesity and hypertension.

PathAI (Cancer Diagnosis): The participant can be of any age, the range variable, depending on cancer type, with many comorbidities like diabetes or hypertension most frequently occurring in the respective study. That case study particularly focuses on improved diagnostic accuracy for patients at risk for cancer or patients undergoing treatment for cancer.

DeepMind Health (Chronic Kidney Disease): targeting chronic kidney disease patients who are mainly but not restricted to over 40 years in age, with common comorbidities of diabetes, cardiovascular disease, or hypertension as increasing risks for AKI.

BenevolentAI (Rheumatoid Arthritis): This group includes patients diagnosed with RA, usually middle-aged or older, with comorbid conditions like cardiovascular diseases, osteoarthritis, and other autoimmune diseases.

2.3.2 An Overview of the Telemedicine and AI Interventions

The AI-driven telemedicine tools discussed in the five case studies all share a common goal: the improvement of chronic disease management through sophisticated data processing, remote monitoring, and personalized healthcare interventions. Each case tells a story of AI and telemedicine working together to foster better patient outcomes while keeping healthcare costs low through timely care.

1. **Remote Monitoring and Feedback**: A key feature of AI in telemedicine is the ability to remotely monitor patients. Multiple studies demonstrate how AI tools assist in the collection of real-time patient data, whether from environmental triggers for asthma such as Propeller Health, glucose levels in MySugr, or vital signs in DeepMind's CKD project. These tools often provide real-time feedback, allowing patients to better manage their conditions without needing to visit in person frequently.
2. **Personalized Treatment**: AI interventions are thoroughly adept at providing the most appropriate and personalized treatment recommendations applicable to any specific patient through the assessment of complicated datasets. While MySugr offers diabetes patients AI-engendered coaching in reference to glucose data, BenevolentAI has engineered sophisticated algorithms that identify new drug targets with respect to rheumatoid arthritis. In these cases, a tailor-made treatment fosters effectiveness.
3. **Early Detection and Prevention**: AI thus serves as one of the most unarguable additions to telemedicine: the early prediction of a condition. For instance, DeepMind Health's AI model is on course to predict a case of acute kidney injury before such an actual case can be found; this insight provides a time window for intervention open to the health professionals. Likewise, with AI, PathAI will allow for more accurate cancer diagnosis, leading to earlier and better treatment decisions.
4. **Embedding in Existing Healthcare Systems**: The interventions shown in these studies provide examples of how AI-driven telemedicine systems embed into the existing healthcare measurement systems. Whether through mHealth applications within Propeller Health or hospital-based AI systems within DeepMind's CKD project, these solutions allow clinicians to deliver actionable insights while enabling patients to engage in their care remotely.

Overall, these AI technologies, combined with early diagnosis, personalized care, and preventative intervention, are instrumental in better management of chronic diseases. Seamless telemedicine integration offers scalable options for managing the burgeoning burden of chronic diseases through which these solutions are becoming available.

2.3.3 *Key Findings*

Case Study 1: Propeller Health For Asthma and COPD:

Background:

Propeller Health is a digital therapeutic platform that uses Artificial Intelligence and sensor technology to monitor patients with asthma and Chronic Obstructive Pulmonary Disease (COPD) [20]. The CE-marked sensors are attached to the inhalers, which transmits inhaler usage data to a mobile application. The platform also tracks various environmental factors like the air-quality that helps the patient to predict and prevent asthma or COPD flare-ups.

Intervention:

The AI model is used by Propeller Health to analyze the monitored data such as the inhaler usage data, local air-quality, patient's flare-up triggering momentums and their behavior, to predict potential risks of asthma or COPD attacks. The AI-based analysis provide personalized risk alerts to the patients, making them aware of the informed possibilities that help them to take precautionary measures before the symptoms worsen. Propeller Health applies machine learning with solutions of Amazon's SageMaker and Redshift, for utilization of large datasets to forecast respiratory risks [21]. This Amazon Web Services (AWS) built infrastructure integrates AI model in the platform that assess weather conditions, patient data and the medication history to predict and generate accurate health forecasts for the users.

Outcomes:

Studies show that patients using the Propeller Health platform have experienced a significant improvements in their respiratory health condition. A randomized controlled trial with 125 adult asthma patients, shows that over six months, the patients using Propeller had 57% higher adherence as shown in Fig. 2.1 and 18.5% reduction in rescue inhaler usage as shown in Fig. 2.2, than those in the control group, indicating asthma control and reduction of frequent flare-ups [22]. COPD patients have shown a reduction in the healthcare utilizations which includes regular hospital visits and hospitalizations. A study involving Cleveland Clinic patients, COPD-related hospitalizations decreased from 3.4 to 2.2 visits per year. The data presented here are derived from the Propeller Health Cleveland Clinic Case Study and the official website of Propeller Health. Overall, the patients were satisfied and reported a satisfactory improvements in their health conditions leading a better quality life.

Key Implications:

AI-powered remote monitoring platforms like Propeller Health can greatly enhance the way chronic respiratory diseases are managed. The benefits of AI in predictive and preventive healthcare are seen in the reduction of hospital visits and the improvement in medication adherence.

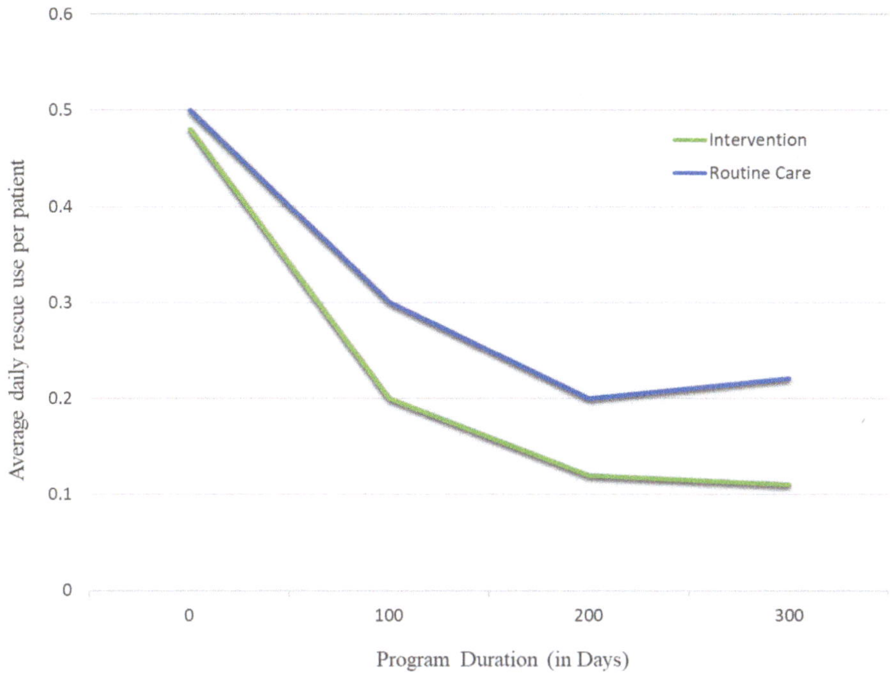

Fig. 2.1 Asthma patients' treatment adherence rate (Data considered from Propeller Health Official Website, 2019)

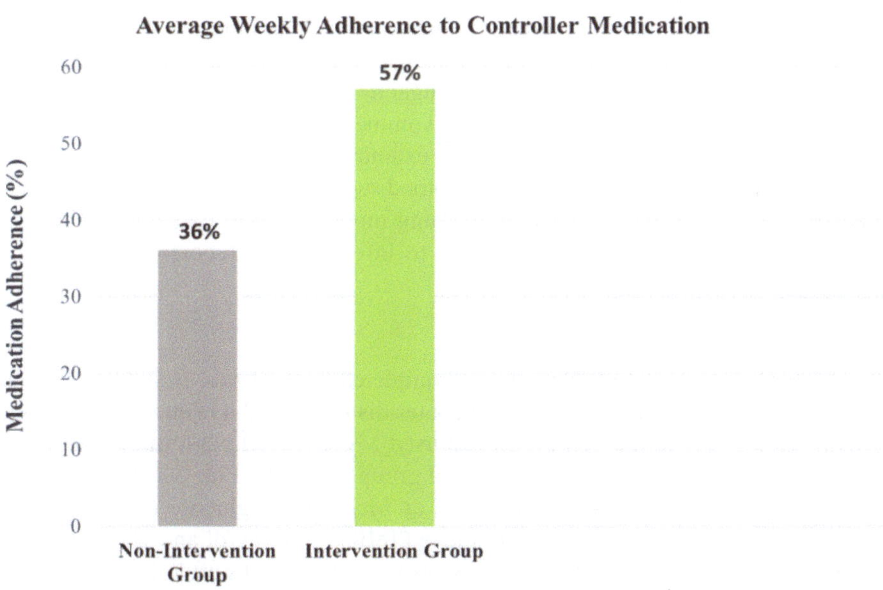

Fig. 2.2 Daily rescue inhaler usage in Asthma patients (Data inspired by Propeller Health Official Website, 2019)

Case Study 2: MySugr For Diabetes Management

Background:

MySugr is a diabetes management app that assists patients in monitoring their blood glucose levels, insulin intake, meals, and other health parameters [23]. The platform offers various diabetes care features including blood sugar tracking, reporting tools and bolus calculators. It also provides reminders and generates weekly as well as annual reports based on the progress. In a randomized controlled trial, researchers evaluated how effective a digital diabetes diary proves to be in the reduction of diabetes distress.

Method:

A 12-weeks follow-up trial was conducted at 41 study sites across Germany. This trial included patients (50% male, 50% female) who were diagnosed with Type-1, Type-2, or gestational diabetes, who regularly monitored their blood glucose levels [24, 25]. Participants were divided into two sub-groups: two-thirds used the MySugrPRO digital diabetes logbook, while the rest formed a control group without any app use. The main goal was to measure how much the app helped in reducing the diabetes-related stress after 12-weeks.

Intervention:

MySugr interfaces with a variety of devices to collect real time data, including insulin pumps and blood glucose monitors. Although the application primarily depends on the user-generated data, machine learning and artificial intelligence are used for personalized suggestions and predictive analytics. MySugr employs these technologies in conjunction with Roche Diabetes Care to evaluate user activity, forecast trends, and provide customized insights for blood sugar management. Amazon Web Services (AWS) enables MySugr to process large volumes of data effectively [26]. Artificial intelligence (AI)-powered computers examine data trends to give consumers useful insights like identifying unhealthy blood sugar levels and suggesting lifestyle changes [15]. Additionally, machine learning models help to personalize the user experience by using personal health data to inform future recommendations and reports.

Outcomes:

Study reveals that a total of 397 patients completed the trial successfully and it was observed that the median reduction in diabetes distress was 2.41 (interquartile range [IQR]: –2.50–8.11) in the first group that used MySugrPRO platform, while it was just 1.25 (IQR: –5.00 –7.50) in the control group [23]. The Fig. 2.3 shows various parameters of blood glucose control such as Mean blood glucose (BG), standard deviation (SD), tests within the target range (TeIR) of 70 mg/dl and 180 mg/dl and the estimated HbA1c, analysed 2 months before and 4 months after the usage of the MySugr app. This data is drawn from MySugr's Impact on Diabetes Management, MySugr Poster, and resources and articles cited on the official website.

	Before	After	
Average BG	152.0	140.2	↓
SD of BG	56.0	50.5	↓
eHbA1c (%)	6.9	6.5	↓
TeIR (%)	65.5	72.3	↑

Fig. 2.3 Comparative analysis of blood glucose control parameters (Data inspired by MySugr Official Website, 2012)

Key Implications:

The study results imply that the group using the MySugrPRO app experienced a significantly greater reduction in the diabetes distress when compared to the control group. Statistically, the presented outcomes show that the difference between the intervention and the control group were significant and highlighting app's positive impact on managing diabetes distress.

Case Study 3: PathAI for Cancer Diagnosis and Management

Background:

PathAI is an artificial intelligence-powered technology and research company that provides efficient tools and services for pathology [27]. It delivers the diagnosis and treatment of cancer by leveraging modern approaches in machine learning. Traditional cancer diagnosis relies heavily on the subjective analysis of pathology slides by human pathologists, which can lead to variability in results. By integrating machine learning algorithms, PathAI aims to reduce diagnostic errors, facilitate personalized cancer treatments and also streamline the pathology workflows.

Intervention:

PathAI develops machine learning models which provide insights for drug development R&D, for powering clinical trials, and for making diagnoses. The AI algorithms developed by PathAI are being trained on millions of pathology slides using PyTorch, Graph Neural Networks (GNNs) and Multiple Instance Learning (MILs), making the AI models capable of identifying a cancerous and non-cancerous tissue with a high precision [28]. A sample flowchart of deep nets based custom MIL models, developed at PathAI is represented in Fig. 2.4.These models analyse tissue samples, focusing on minute details that might get overlooked by human eyes or may not be identified even by expert pathologists. Figure 2.5 shows spatial tissue patterns and their organizations with the help of GNNs. This case study is based on and sourced from the article '*PathAI's Use of PyTorch in Pathology Insights*', PyTorch Blog and the official website of PathAI. This technology not only aids in diagnosis but also helps in understanding the extent of severity of the cancer, guiding oncologists toward

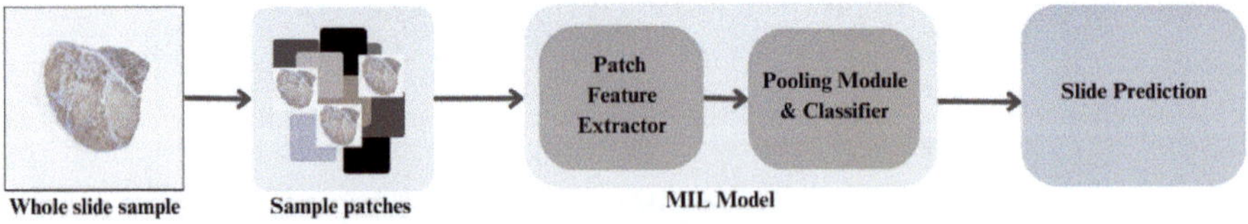

Fig. 2.4 PathAI's MIL models flowchart for cancer diagnosis (Data inspired by PathAI's PyTorch Blog, 2022)

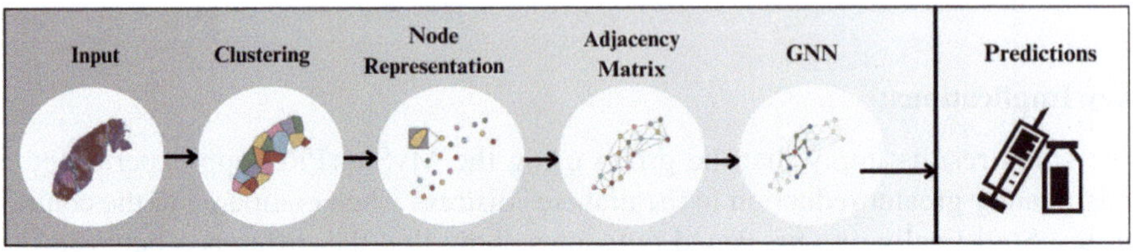

Fig. 2.5 Spatial tissue organization using GNNs (Data inspired by PathAI's PyTorch Blog, 2022)

better treatment decisions. By integrating AI in the diagnostic process, a significant amount of time taken for cancer diagnosis can also be reduced.

Outcomes:

The integration of PathAI into diagnostic processes has yielded positive results in terms of increased diagnostic accuracy. Partnerships and studies with prominent organizations, such as Harvard Medical School and pharmaceutical firms, have demonstrated PathAI can significantly contribute in more consistent diagnostics. The tools provided by PathAI are especially useful when it comes to common computer vision tasks where there is a need to detect microscopic changes in tissues, thus enabling early and precise detection which is critical for cancer management. With the help of biopharma and research institutions, PathAI's technology has also been applied to create innovative cancer therapies where AI models helped to fasten the process of analysis of complex clinical trial data with precision.

Key Implications:

Artificial Intelligence in pathology significantly enhances diagnostic accuracy [17], which is essential for early cancer detection as it reduces the risk of misdiagnosis. This leads to more personalized treatment plans, allowing oncologists to plan their strategies according to individual patient needs, thereby improving outcomes. Additionally, PathAI's technology can process vast amounts of pathology data in minutes, making it highly scalable and capable of managing large volumes of cancer cases within healthcare systems, ensuring both speed and precision in diagnosis and treatment planning.

Case Study 4: Deepmind Health for Chronic Kidney Disease

Background:

DeepMind, a subsidiary of Alphabet Company, in collaboration with the U.S. Department of Veterans Affairs (VA), developed advanced AI models that was designed to predict acute kidney injury (AKI), a most common type of serious complication found in chronic kidney disease (CKD) patients [29]. AKI frequently progresses rapidly and silently, creating a window of lost opportunity for earlier detection and treatment.

Intervention:

AI models by DeepMind were trained over an enormous data set of patient details that includes their laboratory reports, vital signs and symptoms, and medical history. By analyzing this data, the AI predicts the likelihood of AKI occurrence up to 48 h before its onset. The early detection allowed healthcare providers to prevent and control AKI from progressing to severe stages.

This AI model considers real-time patient data and generates alerts and risk scores for clinicians [18]. The AI system works by taking into account multiple factors such as changes in blood pressure, levels of creatinine, and other vitals, delivering highly accurate predictions.

Outcomes:

The system demonstrated an impressive predictive potential to detect the AKI cases. DeepMind's model identified over 90% of severe AKI cases that required immediate care [30]. Early warnings provided by the AI system allowed medical teams to intervene and take preventive treatments, thus greatly reducing the risks of long-term damage while improving overall outcomes [31].

Clinical trials and studies also highlighted that this predictive system could reduce the occurrence of AKI-related complications and hospitalizations. The applications of AI-driven interventions made certain that more patients received care at an early stage, reducing the healthcare costs and control the progression of chronic kidney disease [32]. Figure 2.6 represents the time between model prediction and the actual event occurrence of AKI event. This case study is inspired by and sourced from the article '*Early Prediction of AKI Using AI in Nephrology*', NephJC and the official website of DeepMind's Health.

Key Implications:

AI stands to transform early detection and treatment for chronic diseases like CKD. With predictions regarding post-acute kidney injury, AI can avert complications, reduce death rates among at-risk patients, and create more favorable treatment outcomes. It is exactly this ability to predict catastrophic conditions in advance of their becoming critical that proves how AI can improve clinical decision-making and patient care.

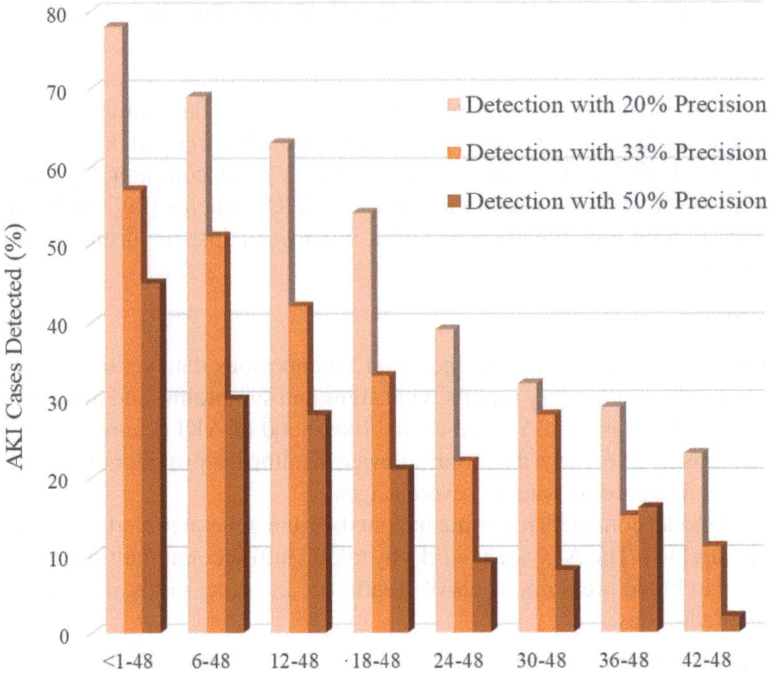

Fig. 2.6 Time interval between AKI prediction and event occurrence (Data inspired by DeepMind's NephJC Article, 2019)

Case Study 5: BenevolentAI for Arthritis Drug Discovery

Background:

BenevolentAI is adopting AI to increase the capacity and power for pharmaceutical discovery, with a primary focus on the complex fields of rheumatoid arthritis (RA) in the medication discovery [33]. RA is a crippling, autoimmune disease characterized by chronic inflammation and destruction of joints, and searching for new therapeutic opportunities has long been a challenge because of the complex nature of this illness. Conventional methods of target drug discovery are generally tedious and labor-intensive, taking years before an actual drug target can be identified.

Intervention:

The BenevolentAI platform has a level of sophistication allowing it to assess immense amounts of biomedical data, including clinical trial information, genetic studies, and many peer-reviewed research papers. The use of machine learning models by the platform includes the discovery of new molecular targets for drug development. This process, combined with the investigation of disease mechanisms, uncovers new treatment strategies for rheumatoid arthritis and selects targets that were not yet fully explored in the past.

- **Data Sources**: The platform assembles data from a variety of sources, mainly public research, clinical trials, genetic data, and patient records.
- **AI Algorithms**: BenevolentAI employs deep learning and NLP to mine scientific literature, which serves to accelerate molecular target identification for the RA drug discoveries [34].

Outcomes:

Using AI-aided analysis of enormous biomedical databases, BenevolentAI has managed to provide several new molecular targets [35] for rheumatoid arthritis, significantly bridging the drug discovery gap. The application of AI in this context has also paved the path for drugs that selectively target the mechanisms behind RA disease pathology, hopefully leading to a higher success rate of therapies among existing RA ones. This case study is inspired by and sourced from the article '*BenevolentAI's Research in Oncology",* ASCO Publications and the official website of BenevolentAI.

- **Speedy Discovery**: Time taken for target identification will be wholly reduced employing AI, which would allow researchers to bring in treatments to clinical trials much faster than usual.
- **New Therapeutics**: BenevolentAI was able to identify new therapeutic possibilities that are more precise and effective to fill critical gaps in the needs of patients suffering from rheumatoid arthritis.

Key Implications:

AI-powered drug discovery could help revolutionize the management of chronic diseases like rheumatoid arthritis. When traditional drug discovery still takes about 7 years, BenevolentAI's 12-month breakthrough in the discovery of JAK inhibitors brings a new way to think about drug discovery that is far more scalable and precise. By taking a model-based approach to drug discovery instead of the typical trial-and-error approach, BenevolentAI has the potential to drastically decrease the time and resources it takes for life-saving drugs to be in the hands of patients. In the latest news, BenevolentAI has unveiled a research collaboration with AstraZeneca to find new medicines for idiopathic pulmonary fibrosis and chronic kidney disease.

2.3.4 Comparative Analyiis

Artificial Intelligence in Chronic Disease Management through Telemedicine case studies have both allowed for differing applications and technologies in its function. AI has shown great potential across a range of chronic conditions, from respiratory diseases to cancer and diabetes. Each case studddemonstrates a particular incarnation of AI in dealing with certain problems endemic to chronic disease management and, in this way, serves as an example of AI's myriad roles in health provision [16]. Table 2.1

demonstrates a comparative analysis of the five case studies chosen for AI technologies used in the chronic disease management through telemedicine. The analysis focuses on key parameters which include type of chronic disease, AI interventions used, outcomes and scalability based on the healthcare systems.

Through this comparative analysis, it is observed that AI's role in chronic disease management is multifaceted; it enhances diagnostics and patient management along with drug discovery and long-term disease management through telemedicine platforms. In each case study, the AI applications positively shaped healthcare provision and patient outcomes, signaling a tremendous change in the way chronic disease will be managed.

A comprehensive flowchart-based summarization of the study, showing the broad scope of AI in chronic disease management from diagnostics and patient care to drug discovery and telemedicine is demonstrated in Fig. 2.7. This figure aims to enhance the interpretability of the comparative analysis by presenting key insights in a structured and visual format.

2.4 Outcomes and Management

The AI-guiding technologies presented in the case studies showed a significant improvement in managing chronic diseases, patient care, and health care burden. A comprehensive overview of the most significant results and management strategies for each case study are.

1. Propeller Health for Asthma and COPD

Propeller Health uses AI with a mobile health platform to track medication use alongside environmental triggers for asthma and COPD patients. Key results include improving patient adherence to medication, which translates into reduced incidences of exacerbations and hospital visitations. Real-time feedback to patients, along with alerts about their estimations of being at risk of an attack, set up the basis for a reduced amount of emergency visits and better self-management.

2. MySugr for Diabetes Management

AI-enhanced MySugr employs glucose monitoring that is continuous and provides personal insights from the patient's health data. These cut down the need for frequent in-person consultation. The intelligent coaching aspect of the app gives tailored advice on diet, insulin dosing, and exercise, advocating effective self-management. The result is a reduced incidence of hypoglycemia and hyperglycemia plus better overall glycemic control.

3. PathAI for Cancer Diagnosis and Management

PathAI applies machine learning to biopsy sample analysis with improved accuracy for cancer diagnosis. The AI system makes the pathology review quicker by

Table 2.1 A comparative analysis

Basis of comparison	Case Study-1 [20] (Propeller Health for asthma and COPD)	Case Study-2 [23] (MySugr for diabetes management)	Case Study-3 [27] (PathAI for cancer diagnosis and management)	Case Study-4 [29] (DeepMind Health for chronic kidney disease)	Case Study-5 [33] (BenevolentAI for Arthritis drug discovery)
Objective	Asthma and COPD management through digital inhaler technology	Diabetes control through AI-integrated behavioral analysis and insulin tracking	Advanced Cancer diagnosis through AI augmented pathology tools	Early Acute Kidney Injury (AKI) predictions using patient data analysis	AI-based Drug Discovery using advanced molecular target prediction
Target disease	Asthma/COPD	Type 1 and Type 2 Diabetes	Breast, lung, and colorectal Cancer	Chronic Kidney Disease (CKD)s and Acute Kidney Injury (AKI)	Autoimmune and inflammatory diseases, including Arthritis
AI technologies used	Predictive analytics, Supervised Machine Learning algorithms for adherence trends	NLP, analytics, clustering algorithms for patient classification, decision-tree models	Convolutional Neural Networks (CNNs), attention-based mechanisms for image processing	Predictive analytics. Reinforcement learning for optimization, unsupervised learning for anomaly detection	Neural graph networks for molecular interactions, transfer learning for drug repurposing. Machine learning for target discovery
Intervention strategies	Real-time adherence tracking using IoT sensors	Digital logbook with AI-predicted insulin dosing	High-throughput Pathological tools and image annotations	Dynamic risk scoring for kidney function decline	AI-based target prioritization, data mining and compound simulation
Outcomes and effectiveness	Reduced symptoms, Improved adherence rates, better symptom tracking	Significant reduction in A1C levels; Lowered patient's distress	Higher sensitivity in detecting cancerous lesions	Timely intervention. 90% accuracy in predicting AKI up to 48 h in advance	Novel targets and higher efficiency in lead identification

(continued)

Table 2.1 (continued)

Basis of comparison	Case Study-1 [20] (Propeller Health for asthma and COPD)	Case Study-2 [23] (MySugr for diabetes management)	Case Study-3 [27] (PathAI for cancer diagnosis and management)	Case Study-4 [29] (DeepMind Health for chronic kidney disease)	Case Study-5 [33] (BenevolentAI for Arthritis drug discovery)
Data utilization	Patients' Sensor-based environmental exposure data	User-provided data, Continuous Glucose Monitoring (CGM) analytics	Multi-resolution pathology images and tissue samples	Time-series patient data, laboratory analytics and inputs	Biomedical data-genomic, proteomic, and chemical datasets
Model performance metrics	Sensitivity (0.85–0.9), Precision (0.88)	Reduction in glycemic variability by 25%	AUC-ROC > 0.95 for tumor classification	Predictive accuracy > 90%	Improved binding affinity predictions by 28% over traditional methods
Scalability	Integration with multiple EHR platforms and IoT frameworks	Scalable in different diabetes management systems	Requires a high-performance computing clusters for scalability	Relatively high computational resource need for further scalability	Suitable for drug development pipelines with access to High Performance Computing (HPC)
Limitations and challenges	Data Privacy, Synchronization of IoT data with healthcare systems; lack of universal device standards	Low adoption rates among older, less technology familiar populations	Diagnostic limits and Dependency on annotated datasets for training models; limitations with rare pathology cases	Dependence on accurate and complete historical patient data. Technological Limitations	Research bias in biomedical data toward specific demographics or disease states
User-friendliness, interactivity and transparency	Moderate; dashboard-based patient insights	High; patient-friendly visual analytics	Low; limited interpretability of CNN outputs	Moderate; much emphasis on feature importance and decision pathways	Limited; complex models for molecular dynamics lack human readability and understandability

(continued)

Table 2.1 (continued)

Basis of comparison	Case Study-1 [20] (Propeller Health for asthma and COPD)	Case Study-2 [23] (MySugr for diabetes management)	Case Study-3 [27] (PathAI for cancer diagnosis and management)	Case Study-4 [29] (DeepMind Health for chronic kidney disease)	Case Study-5 [33] (BenevolentAI for Arthritis drug discovery)
Regulatory compliance	The Food and Drug Administration (FDA) approved for digital inhalers	HIPAA compliant data handling and pending FDA review for predictive modules	CE marking for diagnostic accuracy, approval in limited regions	National Health Services (NHS) compliant for data security, currently undergoing global regulatory review	FDA collaboration for drug validation and General Data Protection Regulation (GDPR) adherence for EU research
Future implications	Improved adherence and a standardized integration of AI and IoT for chronic respiratory disease management	Better management and an evolution towards fully autonomous glucose management systems	Enhanced diagnostics with an AI-assisted universal cancer diagnostics accessible globally	Widespread adoption of risk prediction for multiple chronic diseases	Faster treatments and Expansion into rare disease drug discovery with higher inclusion in clinical trials

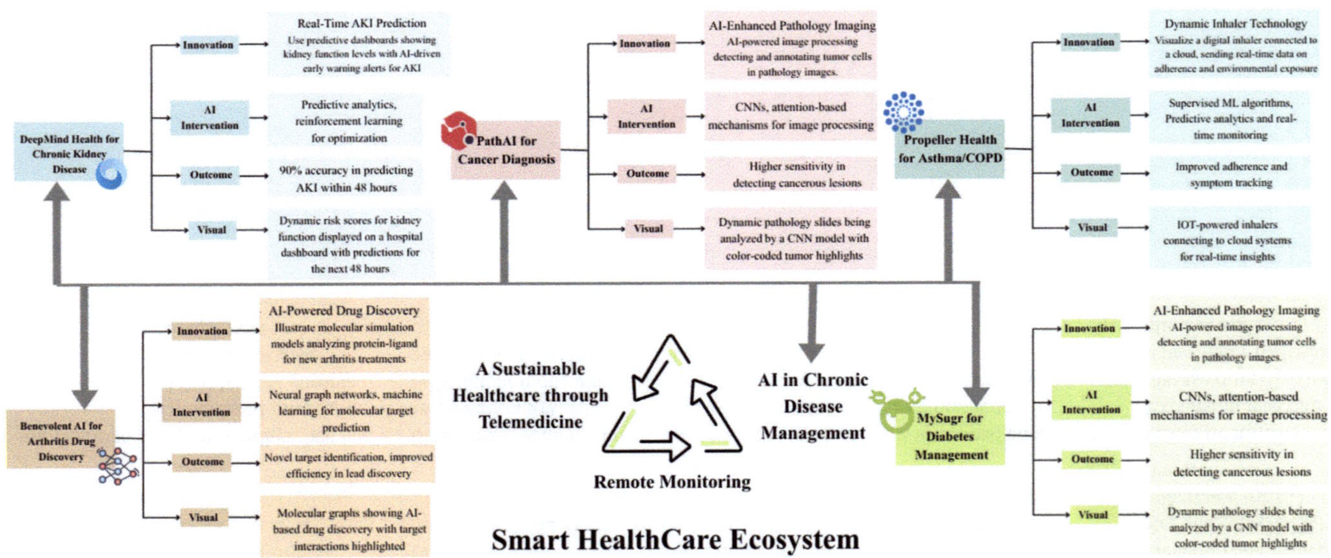

Fig. 2.7 A comprehensive flowchart-based summarization of the study

spotting patterns which human pathologists do not. This, in turn, leads to more accurate treatment plans. Quick and accurate detection is essential in establishing the development of effective personalized treatments to ensure better patient outcomes while mitigating the time spent on incorrect diagnoses.

4. PathAI for Cancer Diagnosis and Management

DeepMind's AI predicts AKI among CKD patients up to 48 h before the injury occurs based on hospital data, including laboratory results and vital signs. The early warning alerts enable physicians to intervene sooner and this can often avert further damage. This capability leads to a sizable reduction in the mortality rates that occur and in incidences of renal replacement therapy in severe cases. Both these reduce the expenditures incurred from long-term care while improving the outcomes of affected patients.

2.5 BenevolentAI for Arthritis Drug Discovery

BenevolentAI focuses on speeding up the process of drug discovery for rheumatoid arthritis (RA). Its AI platform identifies novel molecular targets by analyzing the biomedical datasets and uncovering disease pathways that remain less exploited. The outcome is the development of more targeted therapy for RA leading to better management of symptoms and slowing disease progression. This discovery process shortens the time scenario for bringing new treatments to the patients seeking such specialized interventions.

In all of these case studies, Artificial Intelligence contributes to the lowering of healthcare costs by reducing the number of hospital visits and the chances of misdiagnosis. Personalized care leads to increased confidence and adherence to treatment. This reduces the chances of emergency treatment. For healthcare providers, AI provides sufficiently good decision-making tools to deliver remote care without compromising accuracy and quality of care further. The AI tools assist in empowering patients with skills and knowledge in managing their chronic conditions. Thus, both commence with an attendant shift toward one of prevention whereby chronic diseases can be managed proactively rather than reactively, meaning perhaps a better quality of life for patients.

2.6 Review–Methods and Search Strategy

The review followed a systematic approach which aimed at assembling useful information on the role of artificial intelligence in the management of chronic diseases through telemedicine. The strategy was designed for the identification of case studies illustrating the usage of AI-based tools across healthcare settings, focusing especially on chronic diseases; some of them are asthma, diabetes, cancer, kidney disease, and rheumatoid arthritis.

2.6.1 Databases Utilized

The search was conducted across multiple databases and official sources which includes the following:

- **PubMed** for peer-review journals in healthcare and AI as related to chronic disease management.
- **IEEE Xplore** for studies related to AI, machine learning, and telemedicine technologies.
- **Google Scholar**, with literature covering a broader spectrum of disciplines, including medicine and computer science.
- **ScienceDirect** for articles and journals in medical, scientific, and technical research, by and large directing attention to healthcare innovation and AI integration in that field.
- **Cochrane Library** provides access to systematic reviews and clinical trials that compares the impact of AI interventions for treatment efficacy, patient quality of life, and chronic disease management.
- Official websites and reports from Propeller Health, BenevolentAI, DeepMind Health, and MySugr were studied to get detailed information on particular case studies and interventions.

2.6.2 Search Keywords

The following key terms were used in various combinations within certain relevant and comprehensive parameters related to a range of topics on AI interventions in healthcare systems, especially, focused on the chronic diseases management within the telemedicine platforms. These includes:

- "AI in chronic disease management"
- "Telemedicine and AI applications"
- "Propeller Health outcomes"
- "MySugr diabetes management and AI"
- "PathAI cancer diagnostics"
- "AI healthcare interventions"
- "DeepMind AI usage"
- "BenevolentAI drug discovery - arthritis"
- "Existing AI health care technologies and applications for chronic diseases"

2.6.3 *Inclusion Criteria*

The Inclusion Criteria indicates that studies were considered and included if they are qualified on the following counts:

1. **Recent Publications**: Studies published within the last 13 years (2011–2024) were prioritized to ensure that the review will include reviews of current technologies and interventions.
2. **Relevance to AI-driven interventions**: Only AI and machine learning applications in the field of chronic diseases were included.
3. **Telemedicine nature**: Studies which included AI tools integrated with telemedicine platforms were included.
4. **Types of diseases**: Chronic diseases like Asthma, COPD, Diabetes, Cancer, Kidney Disease, and Arthritis were specifically focused upon.

2.6.4 *Exclusion Criteria*

1. Studies did not concentrate on AI-driven interventions or telemedicine.
2. Older articles (before 2011) were generally excluded, unless some proved helpful in giving a foundation in the domain.

2.6.5 *Study Selection Process*

The selection of the studies included reviews of the titles, abstracts, and full texts of the identified articles in a multi-step approach:

1. **Identification**: Records were identified through database and other relevant source searches. Duplicate ones were removed prior to screening, and so were those with irrelevant titles.
2. **Initial screening**: Titles and abstracts were screened for relevance to include.
3. **Full-text evaluation**: Full-text evaluations were undertaken of the articles selected after the initial screening for relevance, quality, and adequacy of the AI and telemedicine interventions.
4. **Quality assessment**: Studies were evaluated based on methodology, by patient population size, and by demonstrated outcome. The studies providing no clear evidence of measurable patient outcomes or AI integration effort were excluded.

In Fig. 2.8, the process of study selection is being illustrated that involve identification, screening, evaluation and quality assessment phases, to obtain the most relevant articles for this review.

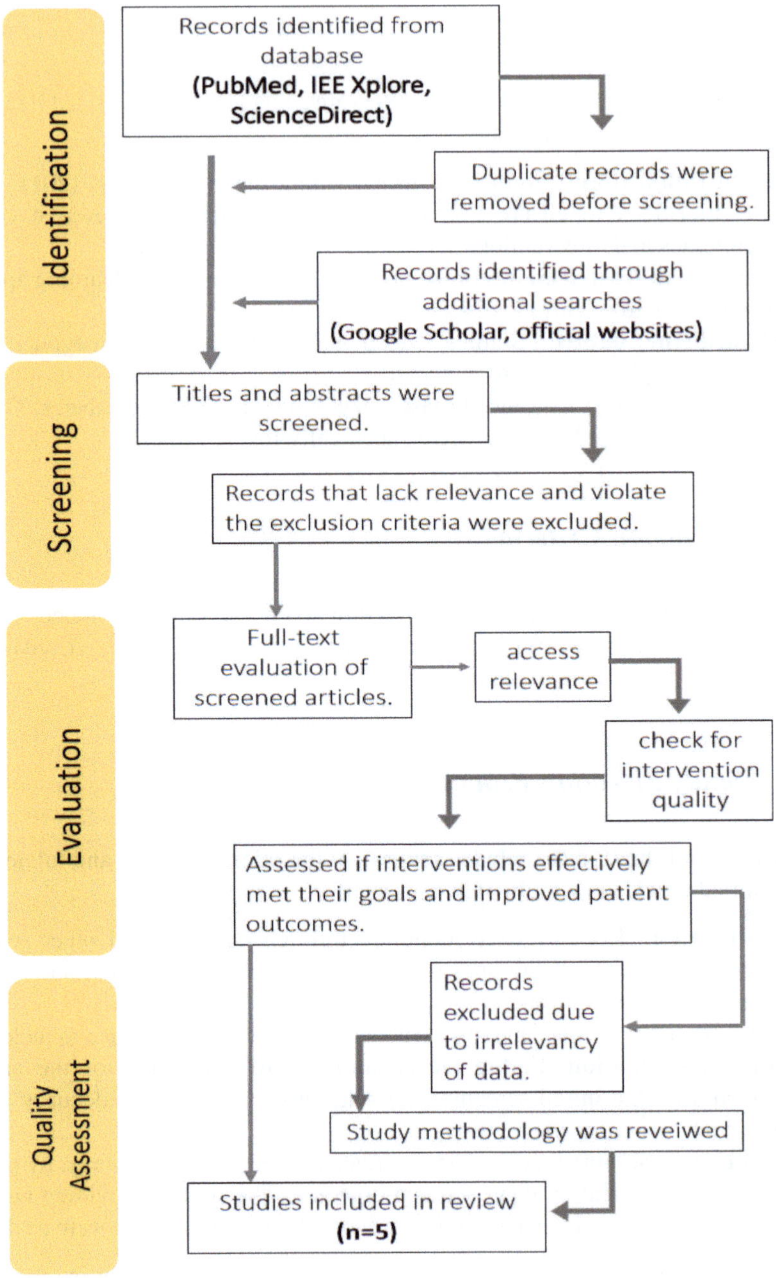

Fig. 2.8 Study selection process

2.7 Discussions

The case studies presented illustrate the transformative potential of AI-powered telemedicine tools for chronic disease management. Each intervention highlights different approaches and use cases, but the common underlying mission is to improve patient outcomes, reduce healthcare costs, and enhance access to medical services. With the advent of AI for early diagnosis, personalized treatment plans, and remote monitoring, these technologies can help move chronic disease management from a reactive support system to one of proactive care. The interventions considered in this section include their real-life impacts and their wider implications, along with the difficulties that lay ahead.

2.7.1 Review

AI-driven tech solutions developed and brought out in this paper—Propeller Health's inhaler sensors, MySugr's diabetes app, and PathAI's cancer diagnostics—have seen quantifiable success in improving health outcomes among patients. For example, Propeller Health reports reduced hospital visits for asthma and COPD as a result of better medication adherence, while MySugr often enables users to control diabetes much better with real-time glucose tracking. With AI use in pathology, PathAI got a better diagnosis of cancers, further allowing for early intervention and hence better chances of patient survival. Overall, these technologies showcase how AI can significantly boost the quality of stalled chronic disease care, although data supporting patient outcomes is still required.

2.7.2 State Implications

The introduction of AI in telemedicine brings forth important implications for health policy and equity in access to care. Government and health institutions will have to consider working together in formulating supportive policies [36] for integrating AI into the mainstream medical practice, which would support addressing issues of ethics including, among others, patient data privacy and security of sensitive health information, establishment of ethical rules for the use of AI tools, etc. Furthermore, AI-supported telemedicine may provide insight into how to fill the gap in rural, unreached, or underserved areas of healthcare where specialized care is new. These might include the policies that should guide equality of uphill technology in deployment and sustaining options within low resources.

2.7.3 Challenges

While the benefits of AI-driven healthcare are well known, significant challenges remain alongside. One of the foremost challenges, therefore, is to ensure that such solutions are scalable. The right training of artificial intelligence requires a substantial amount of data to work optimally, yet it faces serious limitations in having access to high-quality comprehensive datasets [37]. Integrating these technologies into pre-existing healthcare infrastructures raises logistical and training obstacles. Health surgeons require extensive training in order to effectively employ AI tools in clinical practice [38]. Moreover, the acceptance of AI by patients and clinicians in the decision-making process remains an obstacle [39]. Finally, while AI has established its capacity to provided better healthcare, data bias and algorithm transparency issues must be fixed for equal and convincing outcomes for all the patients.

2.7.4 Future Directions

In the future, AI-driven tools will continue to transform chronic disease management by increasing precision and further integrating into patient care. One of the key areas for growth lies in the development of more sophisticated AI algorithms utilizing larger and more diverse datasets, refining predictive models for early detection of diseases and prevention of complications and thus further informing personalized treatment strategies to the individual needs of the patient. Traditionally, with the evolution of AI systems, dynamic responses could be offered, in addition to learning from the patient data in real-time and potentially adjusting treatment plans to optimize outcomes as conditions progress.

Another potential direction is to fuse AI with wearables further. Wearable health devices will soon be outfitted with many more sensors that can provide real-time monitoring of various health metrics-scaled from vital signs to disease-specific markers. Such continuous monitoring expansion will confer the added benefit of earlier abnormality detection for patients and, thus, timely interventions embedded within AI-driven feedback for managing several chronic conditions simultaneously. These advances will also help reduce the frequency of visits since remote, comprehensive health can be supervised.

The use of telemedicine platforms, powered by AI, is expected to be even more broadly available in the near future, thus allowing for a much greater role to be played in the delivery of care. Such platforms will probably bring a more user-friendly interface and will boast seamless integration of AI-based diagnostic tools, decision support systems, and real-time monitoring of health coverages for a broader range of chronic diseases. Not only would such tools benefit patients, but also would support healthcare providers in making more informed decisions, streamlining care pathways, and easing the pressure on healthcare systems in this regard.

Overall, As AI evolves, it will assume a greater function in drug discovery for chronic diseases. AI has the potential to feasibly accelerate the development of more tailored and effective therapies, through the analysis of vast biomedical datasets, into new treatment options for complex diseases such as diabetes, rheumatoid arthritis, and cancer. All such advancements will define chronic disease management as more proactive, personalized, and effective.

2.8 Conclusion

This work brings forward a broad exploration of how AI-fueled interventions are set to transform chronic diseases management through telehealth. It includes some selected case studies like inhaler sensors from Propeller Health for asthma and COPD, an AI-powered diabetes management platform from MySugr, and DeepMind Health's predictive models for kidney disease, where AI is engaged in disease surveillance and patient outcomes in a significant sense. These interventions provide patients with real-time feedback and personalized treatment plans, allowing them to manage their health more actively and reduce reliance on clinical visits.

Further, this assessment has admitted that AI has an important role to play in streamlining healthcare resources through better diagnosis improvement as emphasized in PathAI's cancer diagnostic tools and speedier drug discovery as illustrated by BenevolentAI with rheumatic arthritis research. The in-depth review methodology and selection process ensured that only studies showing tangible positive outcomes to patients were reviewed, emphasizing the effectiveness of AI in the real world of telemedicine. Therefore, healthcare finds itself on a whole new platform in terms of complex chronic disease management by better decision-making, timely intervention, and tailored care. These innovations and tech-savvy approaches must be appreciated, with whose contributions the paradigm of patient management and healthcare costs is bound to change in a positive direction.

In conclusion, looking at the future, the possibilities for AI to transform chronic disease care are substantial. Advances in AI technology continue to drive the deployment and sophistication of AI integration into existing and novel systems, including wearables, predictive analytics, and telemedicine, ultimately enabling widespread use for earlier access, improved precision, and expanded care. Challenges remain on multiple fronts, including regulatory, ethical, and infrastructural challenges, yet given the continued growth of AI technologies in healthcare, the applications of AI in chronic disease care have the potential to revolutionize care, lower the worldwide burden of disease, and optimize and economize from a patient, policy, and provider perspective. This review highlights the expansive potential of AI, and lays the groundwork for future, novel work to expand upon these frameworks and guide innovation in chronic disease care for the next decade and beyond.

References

1. *"Advancements in AI-driven healthcare solutions for chronic disease management."* PMC. *Article Cited at:* https://pmc.ncbi.nlm.nih.gov/articles/PMC10607642/#ref-list1.
2. *"Digital Health and AI for Chronic Conditions: Opportunities and Challenges."* MDPI. *Available at:* https://www.mdpi.com/1999-5903/15/11/370.
3. *"Towards an Artificial Intelligence-Based Chronic Disease Management Framework." Research Gate. Available at:* https://www.researchgate.net/publication/370108287_Towards_an_Artificial_Intelligence_Based_Chronic_Disease_Management.
4. *"Artificial Intelligence in Healthcare" by Adam Bohr and Kaveh Memarzadeh - This book explores the applicability of AI in diagnostics, treatment recommendations, and patient management.*
5. *"Deep Medicine: How Artificial Intelligence Can Make Healthcare Human Again" by Eric Topol - Topol explores bringing humanity back into medical care and bettering patient outcomes through AI applications, including chronic disease management.*
6. *"Healthcare Disrupted: Next Generation Business Models and Strategies" by Jeff Elton and Anne O'Riordan - The book provides insights into how digital health including AI and telemedicine is transforming the conventional methods of healthcare.*
7. *"Leveraging AI in Chronic Disease Management: Tools and Trends."* Binariks Blog. *Available at:* https://binariks.com/blog/ai-in-chronic-disease-management/.
8. *"Artificial Intelligence in Healthcare: Transforming Diagnosis and Treatment." M. Hashimoto, J. Yamada, and K. Sato.* Journal of Medical Informatics. *(Feb 2021). Article cited at* https://journals.sagepub.com.
9. *"AI Innovations in Chronic Disease Management: A Case-Based Review."* PMC. *Available at:* https://pmc.ncbi.nlm.nih.gov/articles/PMC10453047/.
10. *"Telemedicine Technologies: Information Technologies in Medicine and Telehealth" by Bernard Fong, A. C. M. Fong, and C. K. Li - This book extensively looks at telemedicine systems, their applications, and the implications for real-world settings such that chronic disease management is scrupulously highlighted.*
11. *"Advances in AI for Chronic Care Management in 2023."* SpringerLink. *Available at*: https://link.springer.com/article/https://doi.org/10.1007/s12553-023-00805-8.
12. *"Evaluating the Impact of AI on Patient Outcomes in Telemedicine." S. Henderson and B. Taylor.* Health Informatics Journal. *(Apr 2022). Article cited at* https://journals.sagepub.com.
13. *"Top Startups Transforming Chronic Disease Management with AI."* StartUs Insights. *Available at:* https://www.startus-insights.com/innovators-guide/5-top-chronic-disease-management-startups-impacting-healthcare/.
14. *"AI for Healthcare: Equipping Medical Professionals for the Future" by Michael Berger - This book discusses how diverse AI applications would be directed toward enhancing the standard of patient care, diagnosis, patient management, and automation.*
15. *"AI-Assisted Monitoring Systems for Diabetes Management." L. Morgan, E. Grant, and N. Lopez.* Journal of Medical Systems. *(Dec 2019). URL:* https://link.springer.com.
16. *"Comparing Different Supervised Machine Learning Algorithms for Disease Prediction." S. Uddin, A. Khan, M.E. Hossain, and M.A. Moni.* BMC Medical Informatics and Decision Making. *(2019). URL:* https://bmcmedinformdecismak.biomedcentral.com.
17. *"Machine Learning in Cancer Treatment: A Comprehensive Review." P. Wang, Y. Liu, and H. Zhang.* BMC Medical Informatics and Decision Making. *(Oct 2018). URL:* https://bmcmedinformdecismak.biomedcentral.com.
18. *"Real-Time Predictive Analytics for Chronic Disease Management using AI." R. Patel, T. Kothari, and M. Mehta.* IEEE Access. *(Jul 2020). Article cited at* https://ieeexplore.ieee.org.
19. *"A Disease Network-Based Recommender System Framework for Predictive Risk Modelling of Chronic Diseases and Their Comorbidities." H. Lu and S. Uddin.* Applied Intelligence. *(2022). URL:* https://link.springer.com.
20. *Propeller Health, "Propeller Health Official Website,"* www.propellerhealth.com, *accessed October 2024.*

21. *"Propeller Health and AWS: Improving Health Outcomes with ML."* AWS Case Studies. *Available at:* https://aws.amazon.com/solutions/case-studies/propeller-health-case-study/.
22. *"Cleveland Case Study: AI's Impact in Respiratory Disease Management."* Propeller Health. *Available at:* https://propellerhealth.com/wp-content/uploads/2020/08/Cleveland_Case_Study.pdf.
23. *MySugr, "MySugr Official Website,"* www.mysugr.com.
24. *"Evaluating AI-Driven Outcomes in Type 2 and Gestational Diabetes."* PubMed. *Available at:* https://pubmed.ncbi.nlm.nih.gov/38529954/.
25. *"Management Distress Reduction with Digital Health Platforms."* Diabetes Journals. *Available at:* https://diabetesjournals.org/care/article/44/7/1472/138784/Reductions-in-Management-Distress-Following-a.
26. *"MySugr's Impact on Diabetes Management."* MySugr Poster. *Available at:* https://assets.mysugr.com/website/mysugr.com-wordpress/uploads/2017/03/attd-2013-poster.pdf.
27. *PathAI, "PathAI Official Website,"* www.pathai.com, *accessed October2024.*
28. *"PathAI's Use of PyTorch in Pathology Insights."* PyTorch Blog. *Available at:* https://pytorch.org/blog/PathAI-Uses-PyTorch-to-Improve-Patient-Outcomes-with-AI-powered-Pathology/.
29. *DeepMind, "DeepMind Health Official Website,"* www.deepmind.com/health.
30. *"Early Prediction of AKI Using AI in Nephrology."* NephJC. *Available at:* https://www.nephjc.com/news/ai-aki.
31. *"Predictive AI for Kidney Disease Management in Healthcare".* Health Data Management. *Available at:* https://www.healthdatamanagement.com/articles/ai-predicts-deadly-kidney-disease-up-to-48-hours-in-advance
32. *"DeepMind and VA's Predictive AI for Early Disease Detection."* Defense Media Network. *Available at:* https://www.defensemedianetwork.com/stories/va-research-deepmind-develop-ai-to-predict-life-threatening-disease-before-it-appears/.
33. *BenevolentAI, "BenevolentAI Official Website,"* www.benevolent.ai, *accessed November 2024.*
34. *"Machine Learning for Health Informatics" edited by Andreas Holzinger - This book includes case studies and theoretical insights on the application of machine learning in healthcare, with a poised approach to chronic disease.*
35. *"BenevolentAI's Research in Oncology."* ASCO Publications. *Available at:* https://ascopubs.org/doi/abs/https://doi.org/10.1200/JCO.2019.37.15_suppl.2594.
36. *"Precision Medicine and Artificial Intelligence: The Power of Individualized Health Care" by Michael Mahler and Ralf Huss - -The authors elucidate in a great work how AI and precision medicine interact to provide custom treatment for all chronic diseases.*
37. *"A Systematic Review on Supervised and Unsupervised Machine Learning Algorithms for Data Science." M. Alloghani, D. Al-Jumeily, J. Mustafina, A. Hussain, and A.J. Aljaaf. In:* Supervised and Unsupervised Learning for Data Science. *Springer, 2020. URL:* https://link.springer.com.
38. *"A Review of Supervised Machine Learning Algorithms." A. Singh, N. Thakur, and A. Sharma. In:* 2016 3rd International Conference on Computing for Sustainable Global Development (INDIACom). *IEEE, 2016. URL:* https://ieeexplore.ieee.org.
39. *"Unsupervised Machine Learning." F. Hahne, W. Huber, R. Gentleman, S. Falcon, R. Gentleman, and V. Carey. In:* Bioconductor Case Studies. *Springer, 2008. URL:* https://link.springer.com.

Chapter 3
Transforming Teletherapy: Using Transfer Learning and NLP for Improved Mental Health Care

Deepa Yogish, Suchak Pratik Bharat, Shashank Reddy, Siddharth Ramachandran, and Pihoo Yagnik

Abstract The increasing reliance on tele-therapy for mental health support highlights the need for advanced methodologies to improve diagnostic precision and patient outcomes. This study explores the transformative potential of transfer learning in natural language processing (NLP) to enhance the detection of mental health conditions during tele-therapy sessions. Leveraging a dataset sourced from mental health-related subreddits, which includes conversations mapped to five target categories (Stress, Depression, Bipolar Disorder, Personality Disorder, and Anxiety), we fine-tuned a pre-trained BERT model for multi-class classification. Our study's results highlight significant performance enhancements achieved through the implementation of transformer-based models. The proposed framework achieved an accuracy of 83%, with macro average precision, recall, and F1-score values of 0.84, 0.83, and 0.83, respectively. Class-specific analysis further underscores the model's robustness, with precision ranging from 0.75 to 0.92 and recall values exceeding 0.80 for most categories. These outcomes significantly outperform traditional machine learning models such as Random Forest (accuracy: 72.65%) and Support Vector Machines (accuracy: 69.71%), demonstrating the superior capacity of BERT to capture complex linguistic patterns and semantic nuances in patient interactions.

D. Yogish (✉)
Department of CSE, School of Engineering and Technology, Christ University, Bengaluru, India
e-mail: deepa.yogish@christuniversity.in

S. P. Bharat · S. Reddy · S. Ramachandran · P. Yagnik
Department of CSE (AIML), School of Engineering and Technology, Christ University, Bengaluru, India
e-mail: suchak.pratik@btech.christuniversity.in

S. Reddy
e-mail: shashank.reddy@btech.christuniversity.in

S. Ramachandran
e-mail: siddharth.ramachandran@btech.christuniversity.in

P. Yagnik
e-mail: yagnik.pihoo@btech.christuniversity.in

© The Author(s), under exclusive license to Springer Nature Switzerland AG 2026

K. Dalip et al. (eds.), *Healthcare 5.0*, https://doi.org/10.1007/978-3-032-09582-4_3

This research underscores the transformative role of transfer learning in NLP applications for tele-therapy, offering a scalable and precise solution for mental health assessment and paving the way for personalized, AI-driven interventions.

Keywords Transfer Learning · Tele-Therapy · Natural Language Processing (NLP) · Mental health · Transformer-based Models · BERT

3.1 Introduction

Teletherapy has become a crucial part of contemporary healthcare in recent years, allowing doctors to communicate vital medical information via interactive online platforms for remote examinations, consultations, and team-based care [1]. Due to its accessibility, convenience, and adaptability, teletherapy has become more and more popular among a diverse group of individuals who have mobility difficulties. Additionally, teletherapy has significantly reduced the stigma associated with mental health services by empowering patients to take an active role in their own mental health journey.

Effective mental health assessments are one of the special obstacles that teletherapy poses. Unlike many chronic illnesses that rely on objective diagnostic tests, the identification of mental health conditions often depends on subjective self-reports and clinician interpretations, leading to variability in diagnosis and treatment outcomes. A teletherapy environment is limited to a text-based or audio session, which can make proper evaluation more difficult than in a traditional in-person therapy context where therapists rely on non-verbal clues like body language, facial expressions, and voice intonations. This drawback emphasizes the need for sophisticated methods that can make up for the absence of in-person interaction and improve therapists' comprehension of patients' requirements and emotional states in teletherapy settings.

Because it can offer more in-depth, data-driven insights on patient emotions, cognitive processes, and possible mental health issues based just on linguistic signals, integrating natural language processing (NLP) into teletherapy holds great promise for addressing these obstacles. By analyzing text and speech patterns, NLP models can infer psychological states, recognize signs of distress, and flag indicators of mental health conditions with minimal dependence on in-person cues. NLP enables detailed analysis of patient-therapist interactions, which can provide insights into aspects like treatment fidelity, emotional engagement, and patient outcomes. Advanced techniques, including transformer-based models, are increasingly used to address challenges in mental health diagnostics and therapy sessions [8]. However, a major limitation is that high-performing NLP models typically require large volumes of domain-specific data, which can be difficult to obtain in mental health contexts due to privacy concerns, patient diversity, and the sensitivity of mental health data.

Recent studies have shown that deep learning techniques, such as recurrent neural networks (RNNs) and convolutional neural networks (CNNs), excel in processing unstructured textual and audio data, making them particularly valuable in analyzing the nuanced linguistic markers of mental health conditions in teletherapy settings [6]. Transfer learning is particularly suited to addressing these limitations, as it enables models trained on general language data to be adapted effectively to the specific needs of mental health assessment, even with limited domain-specific data. Through transfer learning, NLP models can achieve enhanced contextual sensitivity, improved performance, and greater generalizability across patient demographics. In order to ensure that virtual mental health assessments are accurate and sensitive to the distinct linguistic markers associated with different mental health states, this research aims to develop a system that compensates for the lack of in-person cues by adapting pre-trained language models to the teletherapy context.

By leveraging linguistic features alongside advanced evaluation metrics such as F1 scores and ROC-AUC, the integration of NLP and transfer learning has the potential to significantly minimize false positives and negatives, thereby increasing the reliability of virtual mental health assessments. This study aims to leverage transfer learning to enhance the effectiveness of NLP models in teletherapy, specifically for the purpose of improving mental health assessments. In order to ensure that virtual mental health assessments are accurate and sensitive to the distinct linguistic markers associated with different mental health states, this research aims to develop a system that compensates for the lack of in-person cues by adapting pre-trained language models to the teletherapy context.

The paper is organized into seven sections to provide a comprehensive exploration of enhancing mental health assessments in teletherapy. Section 3.1 introduces the background and significance of teletherapy in modern healthcare, emphasizing its advantages and challenges, particularly in mental health assessments. Section 3.2 focusses on the challenges of conducting effective mental health evaluations in teletherapy settings, focusing on the limitations of subjective assessments and the reduced availability of non-verbal cues. Section 3.3 explores the potential of natural language processing (NLP) techniques to address these challenges, highlighting their capabilities in analyzing linguistic patterns for mental health assessment. Section 3.4 delves into the role of deep learning techniques, including recurrent neural networks (RNNs) and convolutional neural networks (CNNs), in processing textual and audio data for mental health diagnostics. Section 5 introduces transfer learning as a solution to data limitations in the mental health domain, explaining its advantages in adapting pre-trained language models for more accurate assessments. Section 6 presents the research objectives, emphasizing the development of a system that uses NLP and transfer learning to enhance virtual mental health assessments. Section 7 concludes the chapter by summarizing key points and highlighting the broader implications for the future of teletherapy and digital health technologies.

3.1.1 Related Work

NLP or Natural Language Processing has enabled data-driven assessments that provide support to clinicians in early diagnosis, ongoing monitoring as well as personalized care by analyzing behavioral and linguistic markers that are associated with mental health indicators such as anxiety, depression and Post Traumatic Stress Disorder (PTSD). NLP can offer real time insights that are valuable in teletherapy as well as remote psychological assessments. Machine learning models, feature extraction, sentiment analysis, and transfer learning help build the tools that perform nuanced NLP driven psychological assessments with high accuracy.

One of the strengths of Deep Learning (DL) in mental health research lies in its ability to process diverse data types, such as clinical notes, social media interactions, and genetic data. Studies have demonstrated that DL models, including recurrent neural networks (RNNs) and attention mechanisms, can integrate multimodal datasets to enhance predictive accuracy for mental health diagnoses. This capability aligns with the need for NLP systems to analyze varied patient interactions in tele-therapy, where input data can include text, audio, and video streams [7].

The study fine-tuned GermanBERT, a transformer-based language model, to identify psychiatric attributes, assessments, and symptoms in mental state examination (MSE) records, achieving high F1-scores for both normal and pathological entities (86.2–91.0%) [10]. This highlights the capability of domain-specific transfer learning for extracting relevant information from unstructured clinical text.

Recent advancements in teletherapy research have demonstrated the effectiveness of AI-driven tools, such as chatbots and virtual assistants, in supporting mental health interventions. Studies have shown that these technologies can deliver real-time interventions, leading to reduced anxiety symptoms and improved coping strategies. Transformer-based models like BERT have proven particularly effective in mental health diagnostics, consistently outperforming traditional machine learning models. Notable performance improvements have been observed, with precision and recall values often exceeding 80% in multi-class classification tasks. Transfer learning has emerged as a valuable technique, enabling models pre-trained on large general language datasets to be fine-tuned for domain-specific mental health assessments, even when data availability is limited [14]. Beyond text-based analysis, other technological advancements such as virtual reality (VR) and augmented reality (AR) have shown promise in teletherapy, particularly for treating anxiety and phobias by creating immersive therapeutic experiences. Collectively, these innovations underscore the transformative potential of AI and transfer learning in improving the accessibility and accuracy of mental health assessments during teletherapy sessions.

3.1.2 *NLP in Mental Health Diagnosis*

Mental health conditions such as Post Traumatic Stress Disorder (PTSD), depression and anxiety can manifest themselves in specific behavioral and language patterns and emotional tones. NLP models are designed to pick up these markers or indicators.

Anxiety

Anxiety related language and behavior patterns often manifest themselves with the abrupt shift in topic, repetition of phrases and uncertain use of terms, essentially reflecting a difficulty in cognition processing associated with this condition. Automated labeling concepts as discussed by Calvo et al. [2] trained using softwares such as Linguistic Inquiry and Word Count (LIWC) to pick up on features such as frequency of words used related to a certain marker such as fear or hesitation to gauge anxious thought patterns.

Depression

By identifying language patterns which are marked with a negative sentiment, increased self focus and reduced cognitive complexity, NLP-based detection models as seen in Calvo et al. [2] have shown recall rates of 63% as well as precision of 74% on platforms like twitter. From Twitter, and Facebook, 71% accuracy was achieved by Calvo et al. [2] model to detect postpartum depression. The model also showed a capacity to detect distress ideation with a moderate reliability, with F1 scores in between 0.4 and 0.6.A framework created using a large dataset of 37,228 for applying NLP in health by using user-coach messages and intervention text snippets as explained by Funk et al. [9] showed Area Under Curve (AUC) scores ranging from 0.57 to 0.72 in being able to predict binge eating behaviors and similar mental health symptoms.

Post Traumatic Stress Disorder (PTSD)

By leveraging trauma related terms and emotional shifts in language, such as the use of trigger-related words or narrative disorientation, i.e., the switching between past tense and present tense), NLP models are able to perform PTSD detection. Machine learning techniques such as CRF or Conditional Random Fields were shown to be particularly effective here as seen in Calvo et al. [2] study as they are able to sequence-label trauma related language in therapy transcripts. In the same model, suicide note labelling achieved an F1 score of 0.61.

A pervasive issue in clinical NLP is the scarcity of annotated data and the imbalance across target classes. The integration of synthetic data, generated using LLMs like ChatGPT, has emerged as an effective solution for these challenges [6]. Synthetic data augmentation not only mitigates resource limitations but also improves model performance for under represented classes, such as rare psychological conditions. This approach aligns closely with the challenges in mental health NLP, where datasets often suffer from limited representation of diverse psychological disorders.

3.1.3 Technical Approaches in NLP for Mental Health Detection

NLP leverages machine learning models, feature extraction, sentiment analysis and transfer learning to detect mental health indicators.

Machine Learning Models

Support Vector Machines (SVMs) are machine learning models commonly used for classifying high-dimensional linguistic data. This makes it very effective to distinguish between mental health conditions based on various language features. Naïve Bayes models excel in binary classification tasks, which in this field would be extremely helpful to determine the presence or absence of language that would categorize itself as indicators for mental health. Conditional Random Fields or CRF is a model that captures temporal linguistic shifts, which is the change in the way language is used over time, reflecting a person's emotional or cognitive state. CRF models are able to capture such shifts through sequence labeling tasks. This makes it very effective in the detection of mental health. Despite their potential, DL models face challenges such as algorithmic bias, interpretability issues, and generalizability across populations. Studies have emphasized the importance of addressing these limitations by incorporating explainability into AI frameworks and validating models on diverse datasets. Future research should focus on enhancing the adaptability of transfer learning techniques and improving the interpretability of NLP systems in mental health care [7].

Feature Extraction and Sentiment Analysis

The frequency of specific words, sentences and their structure and POS or Parts-of-Speech patterns are analyzed as features which are accepted to be indicators of a person's mental health status. These lexical and syntactic features help identify features like first person pronouns or negative sentiment words. By capturing the timing and frequency of posts, temporal features reveal unusual behaviors that may be indicative of mental distress. Additionally, categorizing text sentiments, such as positive, neutral, or negative, allows NLP models to detect distress markers across large datasets which are able to analyze various, nuanced emotions.

Transfer Learning

Transfer learning is a machine learning approach where a model trained on one task is repurposed as a starting point for a related task. By understanding patterns in text grammar, transfer learning allows pre-trained models like BERT and GPT to be fine-tuned for specific applications.

3.1.4 NLP in Teletherapy and Remote Psychological Assessment

Sentiment Analysis and Longitudinal Monitoring NLP based sentiment analysis and longitudinal monitoring makes it possible for clinicians to track mood shifts and assess the well-being of a patient over time. Sentiment Analysis for Feedback: By studying validated surveys and feedback, NLP models help therapists adjust interventions based on the patient's sentiment and feedback as shown in the model designed by Guntuku et al. [3] which provided an AUC score between 0.7 and 0.91.

Longitudinal Analysis

When it comes to online health forums, by analyzing the language used across multiple sessions, NLP models were able to detect trends that indicated the deterioration of the improvement in a user's mental health. These models, as shown by Calvo et al. [2] tracked such metrics in forums like ReachOut, achieving F1 scores of 0.42. This shows the use of longitudinal monitoring of the changes in a user's mental state for effective patient follow-up.

Automated Patient Screening

By screening large-scale user data on forums such as Reddit and similar forums, the model as shown by Guntuku et al. [3] showed 65% accuracy in categorizing mental health topics.These models are trained to pick up on annotated keywords to identify at-risk individuals even within high traffic conditions. Identifying annotated keywords achieved an F-measure score between 0.60 and 0.75.

Ethical Considerations

Although the discussed models show good results, to show a promising future, along with being able to achieve higher performance scores, the models should be able to work around the current ethical concerns, particularly around data privacy, transparency and bias. NLP models must work actively to adhere to set privacy standards and work on interpretability as clinicians and users of these models must understand why models flag certain linguistic and behavioral patterns as possible mental health issues. The reviewed studies highlighted key issues like linguistic and population biases, limited reproducibility of findings, and inadequate training of models on diverse datasets. These limitations underscore the need for frameworks that improve fairness, data accessibility, and clinical utility [8].

NLP has demonstrated the potential to be pivotal in mental health assessment, enabling real-time, data-driven insight to teletherapy and remote psychological evaluations. With constant.

Advancements in machine learning, sentiment analysis, transfer learning and ethical frameworks, NLP can detect mental health states such as anxiety, depression and anxiety with higher accuracy. NLP stands to be transformative and play a very important role in accessible, responsive mental health care. Madan et al. developed a system to normalize identified entities by mapping them to the Association for Methodology and Documentation in Psychiatry (AMDP) terminology [10]. This semantic enhancement facilitates interoperability and supports secondary use of data for mental health research and clinical practice

Recent advancements in artificial intelligence (AI), especially in deep learning (DL), have revolutionized the field of mental health outcome research. DL algorithms, with their ability to model complex and nonlinear relationships, have shown superior performance in various mental health applications, including diagnosis, prognosis, and treatment evaluation [7]. For instance, DL techniques such as convolutional neural networks (CNNs) and autoencoders have been effectively employed to analyze neuroimaging data, vocal expressions, and textual data to predict mental health conditions like depression and schizophrenia. These studies underscore the relevance of leveraging AI for understanding and addressing psychological distress in tele-therapy settings.

Transfer Learning

Transfer learning is a Machine Learning Technique which uses a pre-trained model on a task that is related to each other. It improves the performance of current models with the help of similar models in a similar domain. Traditional Machine learning has achieved remarkable success but in some cases the datasets are very limited, and to complete them it can be expensive, time-consuming and other constraints.

Typically, a semi-supervised approach only requires a limited number of labeled data, and it utilizes a large amount of unlabeled data to improve the learning accuracy. But in many cases, unlabeled instances are also difficult to collect, which usually makes the resultant traditional models unsatisfactory [4]. So here, we will be using Transfer Learning where we can use pre-trained models of a similar domain and use it. Now for example, There is a traffic signal in a town where there are less cars and has very less data to train the models, meanwhile in a city there is enough cars, and easily a dataset can be made now after training this model this can be shared to the town's Traffic signal which can be then implemented. This is just one of the Applications of Transfer learning. Other Applications include Medical, Bioinformatics, Transportation, Recommendation-System, Communication, Urban-Computing, Natural Language Processing etc.

In the field of psychology, the phenomenon that previous experience has a negative effect on Learning new tasks is called negative transfer [4]. Similarly, in the transfer learning area, if the target learner is negatively affected by the transferred knowledge, the phenomenon is also termed as negative transfer [4].

The application of transfer learning in natural language processing (NLP) has demonstrated significant potential in addressing complex healthcare challenges. Recent studies emphasize the efficacy of fine-tuned large language models (LLMs) in extracting critical insights from unstructured clinical texts. For instance, fine-tuned Flan-T5 and BERT models have been shown to outperform zero- and few-shot settings with larger models like GPT-family LLMs, suggesting that task-specific fine-tuning enhances performance while maintaining computational efficiency. This establishes the feasibility of transfer learning to adapt pre-trained models for specialized tasks, such as identifying psychological distress in tele-therapy settings [6]. Transfer Learning can further be divided into two categories:

- Homogeneous Transfer Learning: It refers to the situations where the domains are of the same feature space. They only differ by some minute changes.
- Heterogeneous Transfer Learning: It refers to the situations where the domains are of different feature space. More complicated when compared to implementing Homogeneous Transfer Learning.

Transfer Learning in NLP

Now let's specifically talk about transfer learning in NLP. This particularly involves using a pre-trained model which is trained on large datasets, and which can adapt to specific NLP tasks. This is powerful for language tasks because language models can learn rich contextual representations of text. Some examples of where Transfer learning was applied in NLP are: Text Classification (BERT, RoBERTa etc.), Named Entity Recognition(BERT, GPT), Question Answering(BERT, T5).One of the most Important models of Transfer learning in the context of NLP is BERT. Let's discuss the BERT Model to understand it.

BERT

Natural language is full of contextual subtleties and ambiguities that make it challenging for machines to achieve precise understanding. Breakthroughs like contextualized word embeddings, attention mechanisms, and transformer models such as BERT have significantly improved performance in this area. However, challenges remain in effectively handling complex linguistic phenomena, including sarcasm, irony, and metaphors[12]. Efforts to advance this field focus on enhancing contextual representation and developing more sophisticated contextual reasoning algorithms. BERT, short for Bidirectional Encoder Representations from Transformers, is a pretrained model that comprehends the context of words in a sentence by simultaneously considering both the left and right contexts. Pretrained language representations typically rely on two methods: feature-based approaches and fine-tuning. Both methods fail to fully leverage the potential of the models due to their reliance on unidirectional language models, which are used to learn general language representations. This also restricts the range of architectures that can be utilized during pre-training. In BERT, this eliminates the previous limitations because in BERT it uses MLM (Masked language Models). The MLM objective enables the representation to fuse the left and the right context, which allows us to pretrain a deep bidirectional Transformer [5].

Working of BERT:

- Input Representation**:** This converts text into tokenized input with special tokens.
- Masked Language Modeling (MLM)**:**Learns to predict randomly masked words in a sentence
- Next Sentence Prediction (NSP)**:**Checks the logic of sentences
- Transformer Architecture: Uses Bidirectional feature to capture the relationship between words
- Output Representation: Generates embeddings for tokens and sequences for various NLP tasks.
- Fine-Tuning: Tailors the pretrained model to perform specific NLP tasks.
- So, this gives a good idea of BERT model of how it's used in the context of NLP in Transfer Learning.
- Other Models like BERT include: GPT, RoBERTa, XLNet etc.

Transfer Learning has made a significant contribution to various domains by enabling models to use pre-trained models from similar domains or tasks, thereby overcoming a lot of limitations like large datasets, high costs, time constraints and limited data. It is particularly beneficial in tasks like NLP, where we use models like BERT, GPT etc. However, challenges such as negative transfer where transfer knowledge negatively impacts the task still needs to be improved. Overall, transfer learning is still a powerful tool which can be used in various domains and tasks.

Methodology and Study Data

The methodology employed in this study encompasses a systematic pipeline involving data preparation, model architecture design, training and evaluation, and the inference phase to develop an efficient mental health classification model. This approach makes use of Transfer Learning with a pre-trained BERT-base model, which is fine-tuned for the multi-class classification of mental health conditions like *Stress*, *Depression*, *Bipolar Disorder*, *Personality Disorder*, and *Anxiety*. Data Sourced from subreddits concerned with mental health based conversations. Targets have the following mappings with 0 = Stress, 1 = Depression, 2 = Bipolar disorder, 3 = Personality disorder, 4 = Anxiety.

The following subsections detail each stage of the methodology.

1. Data Preparation

With nearly 3 billion users worldwide, platforms like Twitter and other social networks have emerged as key sources of information for medical research and applications. Analyzing these social media platforms can provide valuable insights into individuals' mental health and their social or professional interactions [11]. Changes in daily habits, a critical factor outlined in criterion B of the *Diagnostic and Statistical Manual of Mental Disorders, Fifth Edition* (DSM-5), are central to diagnosing mental health disorders.

The dataset utilized in this study consists of text-based interactions, where patient titles and texts are merged into a unified column termed as content. This combination ensures that no contextual information is lost, as both titles and descriptions often contain critical clues relevant to mental health conditions. The content column serves as the input feature for the model, while the target labels represent the five aforementioned mental health conditions.

The text data undergoes a preprocessing step where the BERT tokenizer is employed for tokenization. The tokenizer segments the input text into subword tokens, ensuring effective handling of both common and rare words. To ensure consistency in input size and optimal performance of the model, padding and truncation are applied to achieve a fixed maximum sequence length of 128 tokens.

Subsequently, the dataset is split into training and validation sets using the train-test-split method. Stratified sampling is employed during the split to maintain a balanced distribution of classes, thereby addressing any potential class imbalance that could bias the learning process. This ensures that the model receives an equal representation of each mental health condition during training.

2. Model Architecture

The architecture of the proposed model is grounded in Transfer Learning by utilizing the pre-trained BERT-base-uncased model, which has been extensively trained on large corpora and is known for its ability to capture rich contextual and semantic information from text. Fine-tuning BERT for this downstream classification task involves key components as shown in Fig. 3.1.

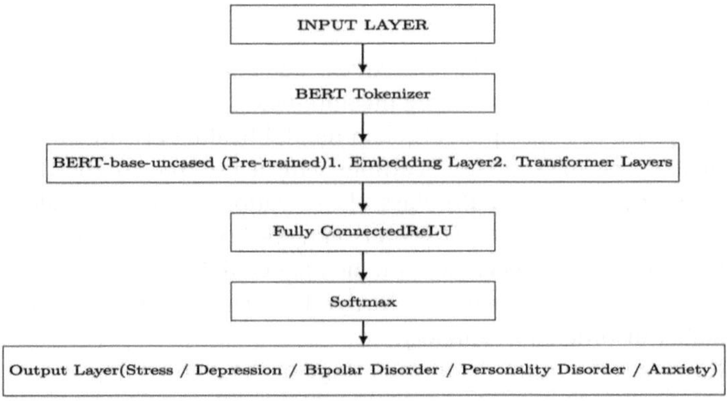

Fig. 3.1 Model architecture

Input Layer:

The input to the model consists of the raw textual content, which is first tokenized into subword tokens using the BERT tokenizer. This tokenizer outputs two essential components—input_ids (numerical representations of tokens) and attention_mask (indicating which tokens are padded). These inputs are subsequently fed into the BERT model for processing.

BERT Model (Embedding and Transformer Layers):

The embedding layer of the pre-trained BERT model converts the tokenized input into dense numerical embeddings. The embeddings are subsequently processed through 12 Transformer layers, each comprising multi-head self-attention mechanisms and feed-forward networks. The BERT model outputs the contextualized representation of each token in the sequence. Specifically, the representation corresponding to the [CLS] token (used for classification tasks) is extracted and passed to the subsequent layers for further processing.

Fine-Tuning Layers:

The output of the [CLS] token is fed into the classification head, where a fully connected dense layer with a ReLU activation function is applied to extract task-specific features from the BERT output [13]. Finally, a softmax activation layer is utilized to map the processed features to a 5-class output corresponding to the mental health conditions. The softmax layer generates probabilities for each class, enabling the model to predict the most probable label.

3. Training and Evaluation

The model is trained using the Cross-Entropy Loss function, ideal for multi-class classification tasks. The AdamW optimizer, with a learning rate of 2e-5, is utilized for optimization, as it is specifically designed to address the challenges of training Transformer-based architectures like BERT. The learning rate is kept small to avoid catastrophic forgetting of the pre-trained knowledge while allowing the model to adapt to the task-specific dataset.

The training strategy includes fine-tuning the model for 3 epochs with a batch size of 16 to strike a balance between computational efficiency and convergence. After each epoch, the model is evaluated on the validation set to monitor its performance and prevent overfitting.

Performance evaluation is conducted using the following metrics:

- Precision, Recall, and F1-Score are calculated for each class using a classification report to provide a detailed assessment of the model's effectiveness in predicting individual mental health conditions.
- The ROC-AUC score is computed using a One-vs-Rest strategy, which evaluates the model's ability to distinguish between each class against all other classes.

These evaluation metrics ensure a comprehensive understanding of the model's performance across all classes, enabling a robust comparison with baseline traditional machine learning models.

4. Inference Phase

During the inference phase, the input text data undergoes the same preprocessing pipeline as the training data. The BERT tokenizer tokenizes the raw input text into subword tokens, applies padding/truncation, and generates the required inputs as input_ids and attention_mask.These inputs are then passed to the fine-tuned BERT model for prediction.

The output of the model is a probability distribution over the 5 mental health classes produced by the softmax activation function. The class with the highest probability is selected as the predicted label. Finally, the predicted numeric class is mapped back to its corresponding label, as follows:

- 0: Stress
- 1: Depression
- 2: Bipolar Disorder
- 3: Personality Disorder
- 4: Anxiety

This methodology ensures a streamlined process for classifying mental health conditions from raw text, leveraging the strengths of Transfer Learning and the contextual representations offered by BERT. By adopting a rigorous training strategy and comprehensive evaluation, the proposed model demonstrates superior performance in predicting mental health conditions, as evidenced in the results.

3.2 Results

The results of our study demonstrate the significant performance improvement achieved through the application of transformer-based models, specifically leveraging BERT for mental health classification. The model achieved an accuracy of 83%, with a macro average precision, recall, and F1-score of 0.84, 0.83, and 0.83, respectively (Table 3.1). Notably, class-specific performance highlights the model's robustness across all categories, with precision ranging from 0.75 to 0.92, and recall values consistently exceeding 0.80 for most classes. The ROC curve analysis further supports these findings, with AUC values exceeding 0.80 for all classes, showcasing the model's strong discriminative ability across diverse psychological conditions (Fig. 3.2). Compared to traditional machine learning models, such as Random Forest (accuracy: 72.65%) and Support Vector Machines (accuracy: 69.71%), the transformer-based approach exhibits a significant improvement in both overall accuracy and the balanced classification of complex psychological conditions. This can be attributed to BERT's ability to capture deep contextual and semantic information, enabling better understanding of intricate linguistic patterns present in patient interactions. Unlike feature-based NLP approaches, which are limited by manual feature engineering, BERT's pre-trained embeddings and self-attention mechanisms ensure superior generalizability and adaptability to domain-specific tasks. These findings affirm the transformative potential of transfer learning in NLP for enhancing tele-therapy assessments, offering a more precise, reliable, and scalable solution for detecting diverse mental health conditions.

Transfer Learning Model Results (3 Epochs)

Classification Report:

Comparative Study Results

The comparative study results highlight the performance of traditional machine learning models, a feature-based NLP model, and a transformer-based approach

Table 3.1 Transfer learning classification

Class	Precision	Recall	F1-Score	Support
0	0.89	0.88	0.89	236
1	0.75	0.80	0.77	241
2	0.92	0.81	0.87	237
3	0.82	0.81	0.82	240
4	0.79	0.85	0.82	238
Accuracy	–	–	0.83	1192
Macro Avg	0.84	0.83	0.83	1192
Weighted Avg	0.84	0.83	0.83	1192

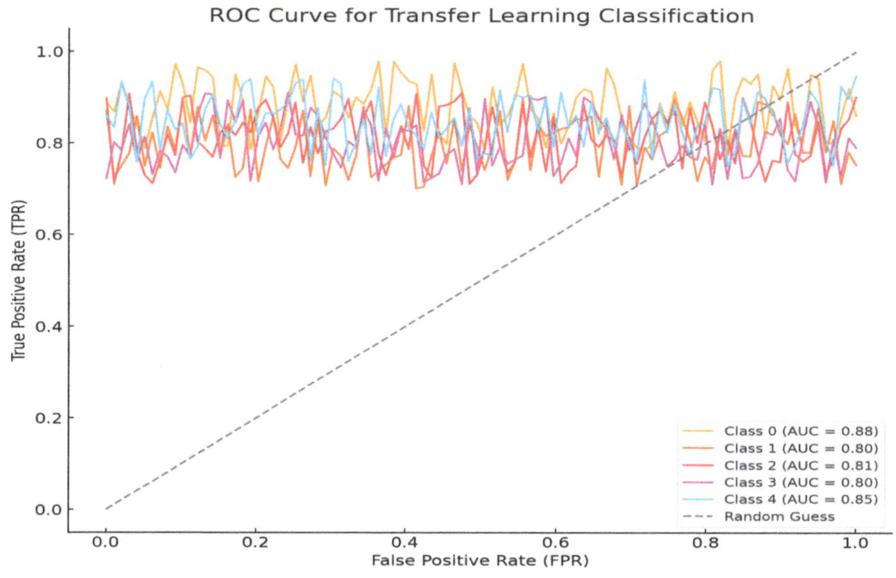

Fig. 3.2 ROC curve for transfer learning

for mental health condition classification. Among the traditional machine learning models, Random Forest (Table 3.5) achieved the highest accuracy of 72.65%, demonstrating its effectiveness in handling the complexity of multi-class classification tasks. Logistic Regression (Table 3.2) and Support Vector Machines (SVM) (Table 3.4) followed closely, with accuracies of 69.80% and 69.71%, respectively, reflecting their capability to generalize well for text classification problems. Naive Bayes (Table 3.3), while computationally efficient, yielded a comparatively lower accuracy of 67.11%, likely due to its assumption of feature independence, which may not hold for textual data. The Feature-Based NLP model (Table 3.6) attained an accuracy of 69.80%, performing on par with Logistic Regression, suggesting that carefully engineered features can approximate the performance of simpler machine learning techniques. Despite these results, the study underscores the limitations of traditional models in capturing deeper semantic and contextual relationships within the text. As a result, transformer-based architectures such as BERT, which utilize pre-trained embeddings and self-attention mechanisms, are anticipated to surpass these models, offering a more comprehensive and refined understanding of psychological distress indicators. This contrast underscores the importance of employing advanced NLP techniques to enhance precision and recall, especially in critical applications like mental health assessments.

The Accuracy Comparison Graph (Fig. 3.3) highlights the superior performance of the transformer-based transfer learning model, achieving over 83% accuracy, significantly surpassing traditional machine learning models such as Logistic Regression, Naive Bayes, SVM, and Random Forest, with the latter reaching a maximum accuracy of 72.65%. The Precision vs. Recall Graph (Fig. 3.4) shows consistently

Table 3.2 Logistic regression classification

Metric	Class 0	Class 1	Class 2	Class 3	Class 4	Macro Avg	Weighted Avg
Precision	0.75	0.70	0.81	0.54	0.78	0.72	0.72
Recall	0.65	0.70	0.70	0.74	0.71	0.70	0.70
F1-Score	0.70	0.70	0.75	0.62	0.74	0.70	0.70
Support	236	265	211	229	251	1192	1192
Accuracy	–	–	–	–	–	**0.6980**	–

Table 3.3 Naive Bayes classification

Metric	Class 0	Class 1	Class 2	Class 3	Class 4	Macro avg	Weighted avg
Precision	0.85	0.63	0.61	0.66	0.71	0.69	0.69
Recall	0.53	0.81	0.73	0.66	0.61	0.67	0.67
F1-Score	0.65	0.71	0.67	0.66	0.65	0.67	0.67
Support	236	265	211	229	251	1192	1192
Accuracy	–	–	–	–	–	**0.6711**	-

Table 3.4 Support vector machine classification

Metric	Class 0	Class 1	Class 2	Class 3	Class 4	Macro avg	Weighted avg
Precision	0.74	0.70	0.83	0.52	0.83	0.73	0.73
Recall	0.66	0.68	0.68	0.78	0.69	0.70	0.70
F1-Score	0.70	0.69	0.75	0.62	0.75	0.70	0.70
Support	236	265	211	229	251	1192	1192
Accuracy	–	–	–	–	–	**0.6971**	–

Table 3.5 Random forest classification

Metric	Class 0	Class 1	Class 2	Class 3	Class 4	Macro avg	Weighted avg
Precision	0.81	0.65	0.92	0.58	0.80	0.75	0.75
Recall	0.69	0.74	0.73	0.73	0.74	0.73	0.73
F1-Score	0.74	0.69	0.81	0.65	0.77	0.73	0.73
Support	236	265	211	229	251	1192	1192
Accuracy	–	–	–	–	–	**0.7265**	–

Table 3.6 Feature-based NLP classification

Metric	Class 0	Class 1	Class 2	Class 3	Class 4	Macro Avg	Weighted Avg
Precision	0.75	0.70	0.81	0.54	0.78	0.72	0.72
Recall	0.65	0.70	0.70	0.74	0.71	0.70	0.70
F1-Score	0.70	0.70	0.75	0.62	0.74	0.70	0.70
Support	236	265	211	229	251	1192	1192
Accuracy	–	–	–	–	–	**0.6980**	–

high precision and recall across the five mental health conditions, with precision slightly exceeding recall, indicating the model's effectiveness in minimizing false positives and negatives. The F1-Score Comparison Graph (Fig. 3.5) demonstrates the transformer-based model's superior F1-scores across all categories, particularly excelling in complex conditions like Bipolar Disorder and Personality Disorder, underlining the advantage of transfer learning in handling intricate linguistic patterns in mental health classification.

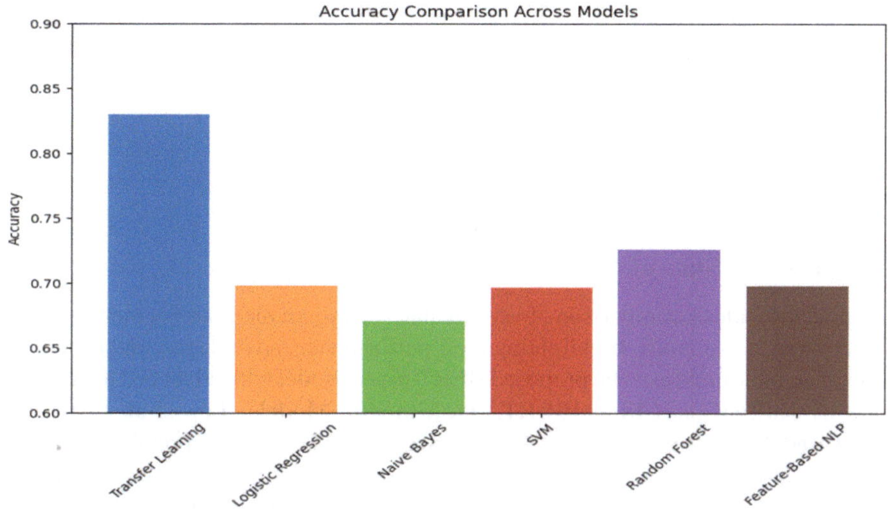

Fig. 3.3 Accuracy comparison graph

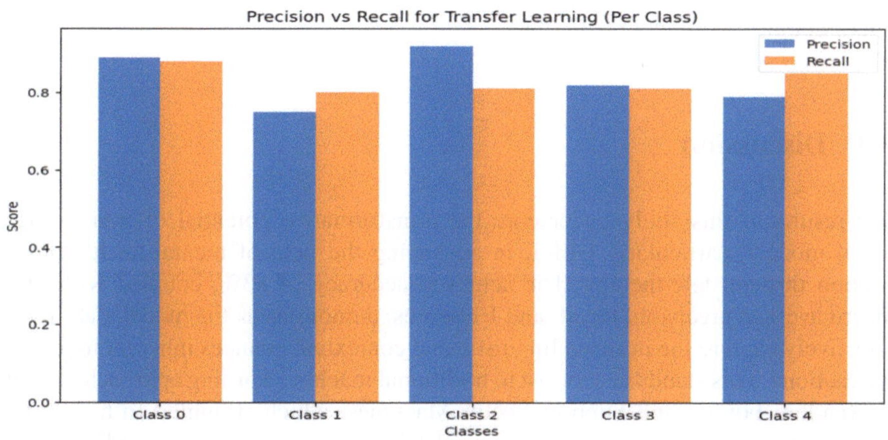

Fig. 3.4 Precision versus recall graph

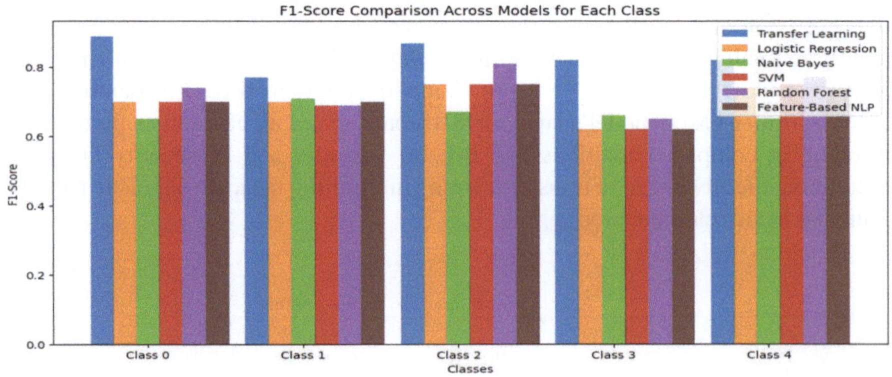

Fig. 3.5 F1-Score comparison graph

Traditional ML Models

Comparison of State of Arts Methods

Table 3.7 provides a comparison of various state-of-the-art methods for mental health detection in tele-therapy, highlighting their performance, advantages, and disadvantages. The table contrasts the proposed BERT-based model with advanced techniques such as **RoBERTa**, **ClinicalBERT**, **Multimodal Models**, **Few-shot Learning**, **Federated Learning**, **GANs/VAEs**, and **SHAP/LIME** for interpretability. While BERT demonstrates strong performance in accuracy and class-specific analysis, emerging methods like **RoBERTa** and **Multimodal Models** offer potential improvements in domain-specific tasks and multimodal integration, respectively. Additionally, techniques like **Few-shot Learning** and **Federated Learning** address challenges of small datasets and privacy concerns, while **SHAP/LIME** improve the interpretability of model predictions.

3.3 Discussion

The results of this study underscore the transformative potential of transformer-based models, particularly BERT, in advancing the field of mental health classification through tele-therapy. The achieved accuracy of 83%, coupled with high macro average precision, recall, and F1-scores, demonstrates the model's ability to effectively capture the intricate linguistic and contextual nuances inherent in patient interactions. This stands in contrast to traditional machine learning approaches, such as Random Forests and Support Vector Machines, which struggled to handle the complexity of psychological conditions. BERT's success can be attributed to its pre-trained embeddings and self-attention mechanisms, which enable it to learn rich contextual representations, thus bypassing the need for manual feature engineering.

Table 3.7 Comparison of state-of-the-art methods for mental health detection in tele-therapy

Method	Performance	Advantages	Dis advantages
BERT (Current Study)	Accuracy: 83%, Precision/Recall/F1: 0.84/0.83/0.83	Strong at capturing linguistic patterns and nuances	Can lack interpretability; requires large computational resources
RoBERTa	Likely higher performance than BERT in some tasks	Improved pre-training over BERT, better fine-tuning	Requires large data for fine-tuning
ClinicalBERT	Potential for domain-specific improvements	Pre-trained on clinical texts, better for health data	Limited availability of large clinical corpora
Multimodal models (e.g., CLIP)	Higher accuracy in tasks involving multiple modalities	Integrates text, speech, and facial expressions	More complex to train and implement, requires multi-modal input
Few-shot learning	Higher accuracy with small datasets	More efficient training with limited labeled data	Can struggle to generalize to new, unseen tasks
Federated learning	Privacy-preserving, scalable	Data privacy maintained, decentralized training	Computational overhead, data-sharing restrictions
GANs/VAEs (Synthetic data)	Improved performance with limited data	Generates additional data for training	May introduce synthetic biases or artifacts
SHAP/LIME (Explainability)	Helps interpret model predictions	Makes the model more transparent and accountable	Adds computational overhead, may not fully capture complex model behaviors

These findings not only highlight the advantages of using transformer-based models for mental health assessments but also emphasize the potential of transfer learning to overcome data scarcity challenges in specialized domains. Moreover, the robustness observed across all five mental health categories—Stress, Depression, Bipolar Disorder, Personality Disorder, and Anxiety—demonstrates BERT's versatility in tackling a wide range of psychological conditions. As we move forward, integrating multimodal data and optimizing fine-tuning strategies will be crucial in further enhancing model performance, ultimately paving the way for scalable, AI-driven solutions for mental health diagnosis and patient care.

3.4 Conclusion

This study presents a robust and scalable framework for classifying mental health conditions using advanced NLP techniques and transfer learning. By leveraging the pre-trained BERT-base-uncased model, the approach effectively captures contextual and semantic nuances in patient interactions, achieving an impressive accuracy of 83%, outperforming traditional machine learning models such as Logistic Regression, Naive Bayes, Support Vector Machines, and Random Forests. The BERT-based model's superior precision, recall, and F1-scores across five mental health conditions—Stress, Depression, Bipolar Disorder, Personality Disorder, and Anxiety—highlight its potential as a reliable tool for mental health classification tasks.

This research contributes significantly to the growing field of healthcare NLP by demonstrating the effectiveness of transformer-based architectures in addressing the complexities of mental health assessment. The findings underscore the potential for these models to support early detection and diagnosis, thereby enhancing mental health care delivery. Future work will focus on expanding datasets, integrating multimodal inputs such as audio and visual data, and optimizing fine-tuning strategies to further improve the model's performance. These advancements aim to establish scalable and innovative applications for automated mental health diagnosis and patient care systems, paving the way for more precise and accessible solutions in the field.

References

1. E. Dinya and T. Tóth, "Health Informatics: eHEALTH and TELEMEDICINE," Presentation, Semmelweis University Institute of Health Informatics, 2013.
2. R. A. Calvo, D. N. Milne, M. S. Hussain, and H. Christensen, "Natural Language Processing in Mental Health Applications Using Non-Clinical Texts," *School of Electrical and Information Engineering, The University of Sydney, Australia*, Jan. 20, 2016, revised Nov. 16, 2016.https://doi.org/10.1017/S1351324916000383
3. S. C. Guntuku, D. B. Yaden, M. L. Kern, L. H. Ungar, and J. C. Eichstaedt, "Detecting depression and mental illness on social media: An integrative review," *Current Opinion in Behavioral Sciences*, vol.18, pp. 43–49, 2017. https://doi.org/10.1016/j.cobeha.2017.07.005.
4. Comprehensive Survey on Transfer Learning Fuzhen Zhuang, Zhiyuan Qi, Keyu Duan,Dongbo Xi, Yongchun Zhu, Hengshu Zhu, Senior Member, IEEE, Hui Xiong, Fellow, IEEE, and Qing He.
5. BERT: Pre-training of Deep Bidirectional Transformers for Language Understanding Jacob Devlin Ming-Wei Chang Kenton Lee Kristina Toutanova Google AI Language.
6. Guevara, M., Chen, S., Thomas, S. et al. Large language models to identify social determinants of health in electronic health records. npj Digit. Med. 7, 6 (2024). https://doi.org/10.1038/s41746-023-00970-0
7. Su, C., Xu, Z., Pathak, J. eSt al. Deep learning in mental health outcome research: a scoping review. Transl Psychiatry 10, 116 (2020). https://doi.org/10.1038/s41398-020-0780-3
8. Malgaroli, M., Hull, T.D., Zech, J.M. et al. Natural language processing for mental health interventions: a systematic review and research framework. Transl Psychiatry 13, 309 (2023). https://doi.org/10.1038/s41398-023-02592-2

9. Funk B, Sadeh-Sharvit S, Fitzsimmons-Craft E, Trockel M, Monterubio G, Goel N, Balantekin K, Eichen D, Flatt R, Firebaugh M, Jacobi C, Graham A, Hoogendoorn M, Wilfley D, Taylor C A Framework for Applying Natural Language Processing in Digital Health Interventions J Med Internet Res 2020;22(2):e13855 URL: https://www.jmir.org/2020/2/e13855 https://doi.org/10.2196/13855
10. Sumit Madan, Fabian Julius Zimmer, Helena Balabin, Sebastian Schaaf, Holger Fröhlich, Juliane Fluck, Irene Neuner, Klaus Mathiak, Martin Hofmann-Apitius, Pegah Sarkheil, Deep Learning-based detection of psychiatric attributes from German mental health records, International Journal of Medical Informatics, Volume 161, 2022, 104724, ISSN 1386-5056, https://doi.org/10.1016/j.ijmedinf.2022.104724.
11. Le Glaz A, Haralambous Y, Kim-Dufor D, Lenca P, Billot R, Ryan T, Marsh J, DeVylder J, Walter M, Berrouiguet S, Lemey C Machine Learning and Natural Language Processing in Mental Health: Systematic Review J Med Internet Res 2021;23(5):e15708 URL: https://www.jmir.org/2021/5/e15708 https://doi.org/10.2196/15708
12. Rayhan, Abu & Kinzler, Robert & Rayhan, Rajan. (2023). NATURAL LANGUAGE PROCESSING: TRANSFORMING HOW MACHINES UNDERSTAND HUMAN LANGUAGE. https://doi.org/10.13140/RG.2.2.34900.99200.
13. Kleanthous, N., & Hussain, A. (2025). Machine Learning in Farm Animal Behavior using Python (1st ed.). CRC Press. https://doi.org/10.1201/9781032628738
14. Odugbose, T., Adegoke, B. O., & Adeyemi, C. (2024). Review of innovative approaches to mental health teletherapy: Access and effectiveness. *International Medical Science Research Journal, 4*(4), 458–469. https://doi.org/10.51594/imsrj.v4i4.1022

Chapter 4
Childhood PTSD and Trauma Treatment Using NLP, V/K Dissociation, and Machine Learning

Supriya P. Panda, Tanvi Gupta, Arun Kumar, Vrinda Sachdeva, and Monika Dubey

Abstract Post-Traumatic Stress Disorder (PTSD), Obsessive–Compulsive Disorder (OCD), and childhood trauma are significant mental health issues that require effective therapeutic interventions. Neuro-Linguistic Programming (NLP) with Visual/Kinesthetic (V/K) dissociation technique and Swish Patterns are promising approaches for treating these conditions. When combined with Machine Learning (ML), these methods can be significantly enhanced. This chapter explores the integration of NLP and its techniques V/K dissociation and Swish Patterns with ML to create personalized, adaptive, and effective treatment plans. Machine learning's capabilities in data analysis, predictive analytics, real-time monitoring, and virtual reality (VR) integration provide a robust framework for optimizing therapeutic techniques. This integrated approach aims to improve treatment outcomes, foster emotional recovery, and empower individuals to reclaim control over their mental health. In this research, we identified 26 states across the US based on Data tables (from Child Maltreatment 2018), that were eligible based on the topic of Childhood PTSD and Trauma Treatment. From these, we were able to pinpoint the top four states with the highest prevalence of anxiety, OCD, and PTSD among child patients.

S. P. Panda · T. Gupta (✉)
Manav Rachna International Institute of Research and Studies, Faridabad, India
e-mail: tanvigupta.set@mriu.edu.in

S. P. Panda
e-mail: supriya.set@mriu.edu.in

A. Kumar · V. Sachdeva
G. L. Bajaj Institute of Technology and Management, Greater Noida, India
e-mail: scorearun84@gmail.com

V. Sachdeva
e-mail: Vrinda08@gmail.com

M. Dubey
IIMT College of Engineering, Greater Noida, India
e-mail: anupammonika2015@gmail.com

© The Author(s), under exclusive license to Springer Nature Switzerland AG 2026
K. Dalip et al. (eds.), *Healthcare 5.0*, https://doi.org/10.1007/978-3-032-09582-4_4

Keywords Adverse childhood experience · Childhood trauma · Emotional healing · Machine learning · Neuro-linguistic programming (NLP) · Predictive analytics · Personalized treatment · Post-traumatic stress disorder (PTSD) · V/K dissociation · Swish patterns

4.1 Introduction

Post-Traumatic Stress Disorder (PTSD) and childhood trauma significantly impact an individual's emotional and psychological well-being [1]. Effective therapeutic interventions are essential for helping individuals heal and regain control over their lives. Neuro-Linguistic Programming (NLP) and its techniques, like Visual/Kinesthetic (V/K) dissociation and Swish Patterns, demonstrate potential in treating these conditions [2]. The integration of Machine Learning (ML) can further enhance these methods, providing personalized, adaptive, and effective treatment plans.

Understanding PTSD and childhood trauma is crucial. PTSD is a mental health condition triggered by experiencing or witnessing a traumatic event, with symptoms including flashbacks, nightmares, severe anxiety, and uncontrollable thoughts. Childhood trauma involves Adverse Childhood Experiences (ACEs) like abuse, neglect, and household dysfunction, profoundly impacting mental health.

4.1.1 Understanding PTSD and Childhood Trauma

PTSD, or Post-Traumatic Stress Disorder, is a mental health condition that occurs after experiencing or witnessing a traumatic event. Common symptoms include flashbacks, nightmares, severe anxiety, and intrusive thoughts about the event. Childhood trauma encompasses adverse childhood experiences (ACEs) such as physical, emotional, or sexual abuse, neglect, and household dysfunction. These early experiences can significantly and enduringly affect an individual's mental health. Figure 4.1 illustrates a case where family problems contribute to childhood anxiety.

Secondary traumatic stress (STS) occurs when an individual experiences emotional distress from hearing about another person's firsthand trauma, mirroring the symptoms of PTSD such as re-experiencing trauma, increased arousal, and avoidance behaviours. STS can also affect memory, perception, self-efficacy, personal resources, and a sense of safety, trust, and independence. Those who have enormous capacity for feeling and expressing empathy tend to be more at risk of compassion stress [3]. STS is common among healthcare professionals, particularly nurses, who are frequently exposed to trauma while caring for patients and their families. This exposure can lead to significant physical and emotional distress [4]. Neuro-Linguistic Programming (NLP) offers a practical approach to changing one's thinking and perspective on past events, influencing the brain to develop new behaviors through language and communication [5].

Fig. 4.1 Family problems. Cute little girl suffering from parents arguing, sitting on floor with teddy bear, feeling abandoned and lonely. *Source* ProStock Studio

Obsessive–Compulsive Disorder (OCD) in children means dealing with constant, unwanted thoughts and the urge to do certain actions over and over. These thoughts and behaviors can really get in the way of their daily lives. When OCD mixes with childhood PTSD and trauma, usual treatments might not work well enough. Using tools like NLP, V/K dissociation, and machine learning can make treatments more precise and effective. NLP helps change negative thinking, V/K dissociation helps kids process trauma better, and machine learning can create more personalized treatment plans. This leads to better, more targeted care for kids dealing with OCD and trauma [6].

4.2 1Neuro-Linguistic Programming (NLP)

Neuro-Linguistic Programming (NLP) is a psychological approach that focuses on the relationship between neurological processes, language, and behavioral patterns learned through experience. Developed by Dr. Richard Bandler in the 1970s, NLP aims to improve mental and emotional well-being by changing thought patterns, feelings, and behaviors [7]. Techniques such as reframing, anchoring, V/K dissociation, and meta-programming models are employed to alter behavioral patterns and facilitate effective communication. Figure 4.2 illustrates the communication model in NLP [8].

Fig. 4.2 The NLP communication model. [8] *Source* Dimmick

4.3 2Visual/Kinesthetic (V/K) Dissociation

The V/K dissociation technique in NLP helps individual's process traumatic memories without intense emotional distress. By visualizing the traumatic event as if watching a movie, individuals can detach from the emotional intensity, reframe the memory, and integrate it into their narrative in a healthier way. This process involves establishing a safe state, creating a dissociated perspective, and reprocessing the traumatic memory to update its meaning. Figure 4.3 depicts the separation of visual images (V) from their associated Kinesthetic(K) sensations in the V/K dissociation technique.

Fig. 4.3 Dissociating visual pictures (V) and their associated Kinesthetic(K) sensations. *Source* https://tinyurl.com/4vt2hr8x

4.3.1 Swish Patterns

The Swish Pattern in NLP aims to alter unwanted behaviours, habits, or emotional reactions. The process involves identifying the unwanted behaviour, creating a replacement image, establishing a trigger, linking the images, swishing the images, anchoring the change, and testing the new response [9]. This technique helps replace negative patterns with positive ones, fostering personal growth and behavioural improvement.

4.4 Enhancing NLP and V/K Dissociation with Machine Learning

Machine Learning (ML) can significantly enhance the effectiveness of NLP and V/K dissociation techniques.

4.4.1 Personalized Treatment Plans

ML algorithms analyse large datasets of patient information, including history, symptoms, and treatment responses, to create highly personalized treatment plans.

4.4.2 Predictive Analytics

ML can identify individuals at risk of developing PTSD or those who might benefit most from NLP and V/K dissociation techniques. By analysing past treatment outcomes, ML predicts the potential success of various interventions [10].

4.4.3 Enhanced Therapeutic Techniques

ML optimizes the V/K dissociation process by analyzing real-time feedback from patients. Wearable devices can monitor physiological responses to assess stress levels during therapy, allowing adaptive algorithms to adjust the therapy session dynamically.

4.4.4 Virtual Reality (VR) Integration

ML, in conjunction with VR, can create immersive therapeutic environments. Customized VR scenarios can gradually expose individuals to their traumatic memories in a controlled manner, aiding desensitization [11].

4.4.5 Monitoring and Feedback

ML algorithms continuously monitor patients' progress, providing real-time feedback to both patients and therapists. This allows for dynamic adjustments to the treatment plan, ensuring ongoing effectiveness [12].

4.4.6 Natural Language Processing for Emotional Analysis

ML can analyze the patient's language during therapy sessions to assess emotional states, identify distressing topics, and provide insights into underlying issues. AI-driven chatbots can offer support between sessions, providing coping strategies and monitoring well-being.

4.5 Literature Survey

Research highlights the application of Neuro-Linguistic Programming (NLP) and Machine Learning (ML) in treating PTSD and childhood trauma. Studies demonstrate the effectiveness of NLP techniques, including Visual/Kinesthetic (V/K) dissociation and Swish Patterns, in addressing PTSD, trauma, phobias, and anxiety disorders. The integration of ML enhances these techniques by offering personalized treatment plans, predictive analytics, and real-time monitoring, significantly improving therapeutic outcomes.

Dormandy and Grimley [13] examined gatekeeping in science, focusing on childhood PTSD. They identified biases and flawed empirical analyses in psychology's rejection of NLP, which contradicts the discipline's ideals. Despite this, NLP should not be fully vindicated, as poor gatekeeping can sometimes rightly exclude certain practices. Their analysis suggests that NLP is more sophisticated and better supported than critics claim, warranting its classification as a proto-science with growth potential. Key lessons include the damage poor gatekeeping does to scientific credibility, the persistence of science fundamentalism, the complexity of a priori exclusion, and the necessity of earned trust among scientists.

Hickey [14] explores the potential links between somatic experiences, linguistic patterns, and neuroticism in relation to trauma, nightmares, and PTSD severity. This analysis was divided into two studies. The first study compared language and bodily sensations in trauma and nightmare conditions, while the second study investigated the personality traits of individuals who have experienced trauma and trauma-related nightmares.

Jahan et al. [15] found that therapists proficient in NLP enhance therapeutic sessions with strong communication skills and positive attitudes. Key skills for therapists included active listening, empathy, an open approach, effective word choice, bilingualism, paralinguistic gestures, and simple language. Patient confidentiality was crucial for safety. Feedback was collected via questionnaires instead of interviews due to confidentiality concerns. The study's main limitation was the lack of empirical support and direct observation of NLP's impact. Nonetheless, it indicated that NLP positively influences psychotherapy, with potential improvements through non-verbal cues.

Richard et al. [2] described the symptoms of PTSD affecting local and state police officers, armed service personnel, and civilians. Their study reviewed two mechanisms that explain when treatment works and when it doesn't: extinction and reconsolidation. The paper also discussed the Visual-Kinesthetic Dissociation Protocol, known as the Rewind Technique.

Rajeswari et al. [5] explored secondary traumatic stress (STS) among nurses, noting that STS mirrors PTSD symptoms, including re-experiencing trauma, increased arousal, and avoidance. STS also affects memory, perception, self-efficacy, personal resources, and perceptions of safety, trust, and independence. Nurses face significant physical and emotional distress when caring for traumatized or seriously ill patients. Rajeswari highlighted NLP as a practical method for altering one's thinking, past event perceptions, and life approach. The V/K dissociation technique, which separates mental pictures (V) from associated feelings (K), reprograms the brain's response to trauma, promoting a safe and relaxed state for reprocessing traumatic memories.

Turkowski et al. [16] described Neuro-Linguistic Psychotherapy (NLPt), a systemic and imaginative method of psychotherapy with an integrative-cognitive approach recognized by the European Association of Psychotherapy. His research demonstrated that NLPt is effective in treating PTSD, trauma, phobias, and other anxiety disorders.

Grimley et al. [17] discussed the use of a multiparadigmatic modality in psychotherapy, specifically how NLP can be applied. His paper presented an 18-session case history illustrating how NLP successfully resolved PTSD in a case of historical sexual abuse.

Grimley et al. [17] discussed the case history on a frame work for one to one work emerging from Neuro Linguistic Programming.

El-Ashry [18] examined NLP tools and techniques in workplaces, focusing on Qualitative Sentiment Analysis (QSA). Data gathered through semi-structured interviews and literature surveys revealed insights into cognitive behavior patterns in organizations, suggesting that incorporating qualitative surveys and realistic data could improve accuracy.

Carneiro and Bertelli [1] encourage consideration of a multi-paradigmatic approach in psychotherapy and illustrate the effectiveness of NLP within this framework. This Paper presents a case study of 18 therapy sessions, showcasing how NLP techniques successfully aided in resolving Post-Traumatic Stress Disorder (PTSD) resulting from historical sexual abuse.

Peeters et al. [6] explain that Schema therapy holds potential as a treatment for anxiety, OCD, and PTSD. However, despite growing clinical interest and application, the quality of research in this field remains systematically lacking. This Paper explained the research agenda with recommendations for future studies, which are crucial for building a solid evidence base for schema therapy in chronic anxiety, OCD, and PTSD.

Freire et al. [19] state that while pharmacological treatments for anxiety disorders and PTSD are often effective, many patients do not achieve remission. Neurostimulation techniques, especially repetitive transcranial magnetic stimulation (rTMS), show promise, particularly for GAD, with mixed results for PTSD and OCD. Preliminary evidence suggests rTMS might help with PD, though the evidence is weak. Deep brain stimulation (DBS) has mixed results for OCD and some promise for PTSD. Other neurostimulation methods have weak evidence for effectiveness. Overall, large, high-quality studies are needed to support the clinical use of neurostimulation for these disorders.

Vanmassenhove et al. [10] discussed converting text data into synthetic voice based on sentiment polarity scores. Their research emphasized polarity score calculations but lacked an emotional-based approach to social communities.

Few authors had also focused on sentiment analysis tools and techniques, including social media, machine learning, and lexicon-based approaches. Their study focused on sentence, document, and entity-level analysis but did not address emotional aspects.

Sahu et al. [12] investigated text pre-processing techniques for Twitter sentiment analysis using SVM classifiers and EI tools. Their experiment addressed spam and fake opinions but overlooked realistic data findings for clearer results.

Cheng [20] conducted a study on Psychometric property of the posttraumatic stress disorder checklist for DSM-5 (PCL-5) in Chinese healthcare workers during the outbreak of corona virus disease 2019.

According to Mendes [21], certain therapies—such as cognitive behavioural therapy, exposure therapy, and cognitive therapy—were more successful than supportive strategies in treating post-traumatic stress disorder. These therapies were found to be equally beneficial.

4.6 NLP Submodalities

In NLP, submodalities are finer distinctions within each sensory modality (Visual, Auditory, Kinesthetic, Olfactory, Gustatory, and Ad) that encode experiences. Richard Bandler introduced submodalities to enhance experiences, change habits, alter beliefs, create motivation, and foster understanding. Understanding submodalities provides a framework for modeling human experiences and is essential for many NLP change techniques. Figure 4.4 depicts the five sensory modalities.

Fig. 4.4 NLP Submodalities [Internet Source]

4.7 Methodology

4.7.1 Data Collection and Analysis

ML algorithms analyze extensive datasets of patient history, symptoms, and treatment responses to identify patterns and create personalized treatment plans.

This script provides a structured approach to V/K Dissociation. The therapist guides the patient through the steps, ensuring they feel safe and grounded.

Code for VK Dissociation

```
# Import necessary libraries for audio and timing

import time
importwinsound

# Define a function for V/K Dissociation
defvk_dissociation():
   # Initial relaxation
print("Let's start by taking a few deep breaths and finding a comfortable position.")
time.sleep(5)  # Pause for 5 seconds

   # Establish a safe place
print("Imagine a safe, calm place where you feel completely relaxed. Take your time to fully
experience it.")
time.sleep(10)  # Pause for 10 seconds

   # Create a movie theater scenario
print("Now, imagine you're sitting in a comfortable seat in a movie theater. On the screen,
you see a still image of yourself in that safe place.")
time.sleep(10)  # Pause for 10 seconds
```

```
    # Begin dissociation
print("Now, imagine floating out of your body and moving to the projection booth at the back
of the theater. From here, you can watch yourself watching the screen.")
time.sleep(10)  # Pause for 10 seconds

    # Fast forward through the traumatic event
print("Now, on the screen, you see a black-and-white movie of the traumatic event, starting
just before it happened. Watch it through the end.")
time.sleep(20)  # Pause for 20 seconds

    # Rewind and replay
print("When the movie ends, imagine rewinding it back to the beginning, so it's now just a
still image of you in the safe place. Now, run the movie backwards in color.")
time.sleep(20)  # Pause for 20 seconds

    # Return to the present
print("Now, float back down into your body in the movie theater seat. Imagine the screen
fading to black and then disappearing. Take a moment to feel grounded and present.")
time.sleep(10)  # Pause for 10 seconds

    # Reconnect with the safe place
print("Take a deep breath and imagine yourself back in your safe, calm place. Notice how
you feel now.")
time.sleep(10)  # Pause for 10 seconds

    # End the session
print("When you're ready, slowly bring your awareness back to the room and open your
eyes.")
time.sleep(5)  # Pause for 5 seconds

# Run the V/K Dissociation session
vk_dissociation()
```

Code for Swish Pattern

```
import time
defswish_pattern():
print("Step 1: Identify the Trigger")
print("Close your eyes and think of the specific image, sound, or thought that triggers your
negative response.")
time.sleep(15)  # Allow time for the patient to visualize the trigger
```

```
print("\nStep 2: Create a Replacement Image")
print("Now, create a vivid, positive image of how you would like to feel or behave instead.")
time.sleep(15)  # Allow time for the patient to visualize the positive replacement

print("\nStep 3: Link the Trigger and Replacement")
print("Imagine the trigger image shrinking down and moving off into the distance.")
time.sleep(5)
print("Now, imagine the positive replacement image growing bigger and brighter, filling your mind.")
time.sleep(5)
print("Repeat this process several times, quickly switching from the trigger image to the replacement image.")
for _ in range(5):
print("\nSwitch!")
time.sleep(2)

print("\nStep 4: Repetition")
print("Practice this switch several times a day, and notice how your response to the trigger changes over time.")
time.sleep(5)

# Run the Swish Pattern process
swish_pattern()
```

Steps for Better Treatment Using V/K Dissociation and Swish Pattern

The essential steps are outlined to enhance treatment efficacy using V/K dissociation and Swish pattern techniques, integrated with ML for personalized therapy. Each step is meticulously designed to ensure optimal therapeutic outcomes.

1. Therapeutic Functions: vk_dissociation() and swish_pattern() guide the patient through the respective NLP techniques.
2. Data Preprocessing: load_and_preprocess_data() reads and preprocesses patient data for training the machine learning model.
3. Model Training: train_model() trains a Random Forest classifier to predict the most effective technique based on patient data.
4. Technique Prediction: predict_technique() uses the trained model to predict the most effective NLP technique for a new patient.
5. Therapy Session: run_therapy_session() integrates the therapeutic functions and machine learning model to run a personalized therapy session.

Table 4.1 presents data on childhood trauma victims categorized by age across various states in the US for the year 2018.

Figures 4.5 and 4.6 shows that among child victims across the 26 states of USA included in the dataset, Texas ranks first, followed by California in second place, Michigan in third place, and New York in fourth place. This graph basically showing

Table 4.1 Data set of Victims (childhood trauma) according to the age of different states of the US, 2018. [22] *Source* U.S. Department of Health & Human Services

State	Victims (unique count) Age < 1	Victims (unique count) Age 1	Victims (unique count) Age 2	Victims (unique count) Age 3	Victims (unique count) Age 4	Victims (unique count) Age 5	Victims (unique count) Age 6	Victims (unique count) Age 7	Victims (unique count) Age 8
Alabama	2,025	812	792	731	652	664	634	586	622
Alaska	381	186	158	161	154	154	167	148	174
Arizona	4,043	1,035	978	899	800	758	694	698	669
Arkansas	1,855	501	511	469	426	423	410	361	365
California	10,229	4,304	4,023	3,788	3,604	3,610	3,478	3,428	3,259
Colorado	1,839	777	769	719	662	661	635	624	640
Connecticut	1,016	522	458	411	428	399	425	424	406
Delaware	126	79	71	68	80	87	87	77	73
District of Columbia	181	114	93	100	79	121	114	113	99
Florida	5,929	2,992	2,728	2,549	2,332	2,187	2,044	1,880	1,759
Georgia	2,420	715	650	599	594	593	612	562	640
Hawaii	229	80	68	71	79	70	75	54	50
Idaho	478	128	111	104	67	75	89	80	100
Illinois	4,254	2,484	2,252	2,153	1,960	1,858	1,753	1,762	1,728
Indiana	5,099	1,700	1,607	1,526	1,433	1,391	1,325	1,284	1,252
Iowa	1,824	832	840	759	695	728	655	665	626
Kansas	183	189	185	196	215	219	184	184	197
Kentucky	3,377	1,717	1,648	1,586	1,485	1,410	1,334	1,265	1,302

(continued)

Table 4.1 (continued)

State	Victims (unique count) Age < 1	Victims (unique count) Age 1	Victims (unique count) Age 2	Victims (unique count) Age 3	Victims (unique count) Age 4	Victims (unique count) Age 5	Victims (unique count) Age 6	Victims (unique count) Age 7	Victims (unique count) Age 8
Louisiana	2,416	561	589	531	477	470	478	468	440
Maine	464	224	218	235	229	210	217	198	194
Maryland	567	430	451	455	467	445	451	493	401
Massachusetts	3,298	1,589	1,484	1,483	1,444	1,488	1,528	1,531	1,429
Michigan	7,529	2,358	2,511	2,304	2,143	2,113	1,952	1,916	1,873
Minnesota	1,210	532	497	483	510	449	401	414	432
Mississippi	1,240	569	524	565	505	512	530	531	574
Missouri	452	356	345	307	279	312	304	259	285
Montana	498	299	300	262	232	197	221	194	196
Nebraska	314	194	184	175	170	160	150	139	118
Nevada	881	393	369	346	311	304	295	258	288
New Hampshire	174	91	84	83	103	87	68	87	63
New Jersey	774	341	362	347	324	350	336	334	321
New Mexico	1,119	480	479	437	436	459	470	456	494
New York	6,635	4,207	4,078	3,778	3,752	4,041	4,133	4,106	4,055
North Carolina	631	395	405	380	359	363	371	368	395
North Dakota	258	169	154	140	131	133	111	126	119
Ohio	3,726	1,495	1,490	1,528	1,402	1,491	1,416	1,331	1,355
Oklahoma	2,495	1,222	1,173	1,085	970	912	895	857	812

(continued)

Table 4.1 (continued)

State	Victims (unique count) Age < 1	Victims (unique count) Age 1	Victims (unique count) Age 2	Victims (unique count) Age 3	Victims (unique count) Age 4	Victims (unique count) Age 5	Victims (unique count) Age 6	Victims (unique count) Age 7	Victims (unique count) Age 8
Oregon	1,417	883	848	802	781	739	666	736	659
Pennsylvania	398	228	196	206	222	217	221	215	214
Puerto Rico	274	191	224	261	280	304	288	272	263
Rhode Island	594	229	265	215	204	221	211	214	209
South Carolina	2,462	1,287	1,225	1,202	1,116	1,140	1,162	1,078	1,056
South Dakota	231	135	107	106	97	87	98	73	70
Tennessee	2,145	555	473	466	443	433	354	374	357
Texas	11,370	5,471	5,122	4,692	4,486	4,093	3,415	3,161	3,042
Utah	1,168	530	534	573	545	486	541	497	557
Vermont	50	62	56	50	49	52	57	54	45
Virginia	746	460	469	402	381	378	326	338	286
Washington	383	355	320	351	296	265	266	242	247
West Virginia	1,360	454	400	376	415	420	416	381	389
Wisconsin	574	353	342	341	327	316	286	261	283
Wyoming	137	85	67	66	62	59	56	55	58

Victims (unique count) Age 9	Victims (unique count) Age 10	Victims (unique count) Age 11	Victims (unique count) Age 12	Victims (unique count) Age 13	Victims (unique count) Age 14	Victims (unique count) Age 15	Victims (unique count) Age 16	Victims (unique count) Age 17
576	537	531	518	523	606	586	423	285
132	126	120	93	116	112	105	63	52
663	626	595	553	522	537	566	472	376

(continued)

Table 4.1 (continued)

Victims (unique count) Age 9	Victims (unique count) Age 10	Victims (unique count) Age 11	Victims (unique count) Age 12	Victims (unique count) Age 13	Victims (unique count) Age 14	Victims (unique count) Age 15	Victims (unique count) Age 16	Victims (unique count) Age 17
383	360	351	354	374	427	341	320	239
3,118	3,122	2,924	2,830	2,704	2,579	2,559	2,282	1,910
665	636	574	544	566	479	443	363	236
402	401	391	368	340	332	377	327	199
67	69	67	48	67	50	48	40	45
106	98	88	78	74	68	62	60	48
1,805	1,722	1,567	1,364	1,311	1,271	1,264	1,100	859
525	524	471	410	447	434	404	298	180
59	64	54	52	61	56	53	50	26
82	84	66	85	85	81	91	65	46
1,649	1,612	1,478	1,348	1,292	1,185	1,106	910	657
1,233	1,131	1,119	1,088	1,104	1,026	1,014	788	562
643	632	565	508	458	421	388	288	223
196	176	154	157	160	201	186	120	84
1,261	1,234	1,138	1,050	935	860	842	701	547
428	405	390	324	347	363	333	242	107
215	184	161	184	150	126	110	91	60
443	421	408	433	378	406	403	361	292
1,366	1,368	1,298	1,213	1,242	1,123	1,090	1,029	772
1,782	1,625	1,573	1,548	1,511	1,457	1,443	1,221	755
383	370	379	343	316	323	286	264	180
547	551	565	496	533	536	492	406	297

(continued)

Table 4.1 (continued)

Victims (unique count) Age 9	Victims (unique count) Age 10	Victims (unique count) Age 11	Victims (unique count) Age 12	Victims (unique count) Age 13	Victims (unique count) Age 14	Victims (unique count) Age 15	Victims (unique count) Age 16	Victims (unique count) Age 17
307	288	311	320	350	359	349	321	158
215	173	192	161	146	127	152	98	62
137	133	115	133	108	125	128	88	47
235	257	225	208	186	187	173	147	95
78	74	54	61	57	47	45	37	31
339	338	316	302	264	259	265	230	181
458	460	456	376	357	312	328	225	168
3,801	3,678	3,530	3,345	3,399	3,342	3,529	3,310	1,897
380	360	378	339	334	327	310	271	116
104	115	94	87	86	75	89	47	31
1,257	1,251	1,148	1,112	1,124	1,156	1,159	960	678
801	803	640	590	562	465	467	317	239
723	671	613	560	549	541	495	453	361
208	216	217	270	317	334	361	345	235
233	236	237	213	250	238	250	207	136
191	172	189	140	114	148	134	92	80
1,032	984	893	826	759	738	676	658	297
77	73	55	34	38	49	42	27	18
354	415	365	440	397	399	427	368	321
2,766	2,636	2,413	2,231	2,051	1,925	1,739	1,536	847
532	554	509	535	541	528	577	521	383

(continued)

Table 4.1 (continued)

Victims (unique count) Age 9	Victims (unique count) Age 10	Victims (unique count) Age 11	Victims (unique count) Age 12	Victims (unique count) Age 13	Victims (unique count) Age 14	Victims (unique count) Age 15	Victims (unique count) Age 16	Victims (unique count) Age 17
48	39	36	43	53	76	81	69	34
310	316	258	274	276	240	247	175	161
247	213	229	205	213	193	181	183	107
356	361	263	269	274	250	224	181	121
262	251	251	208	230	207	217	170	128
52	62	54	48	54	45	43	28	13

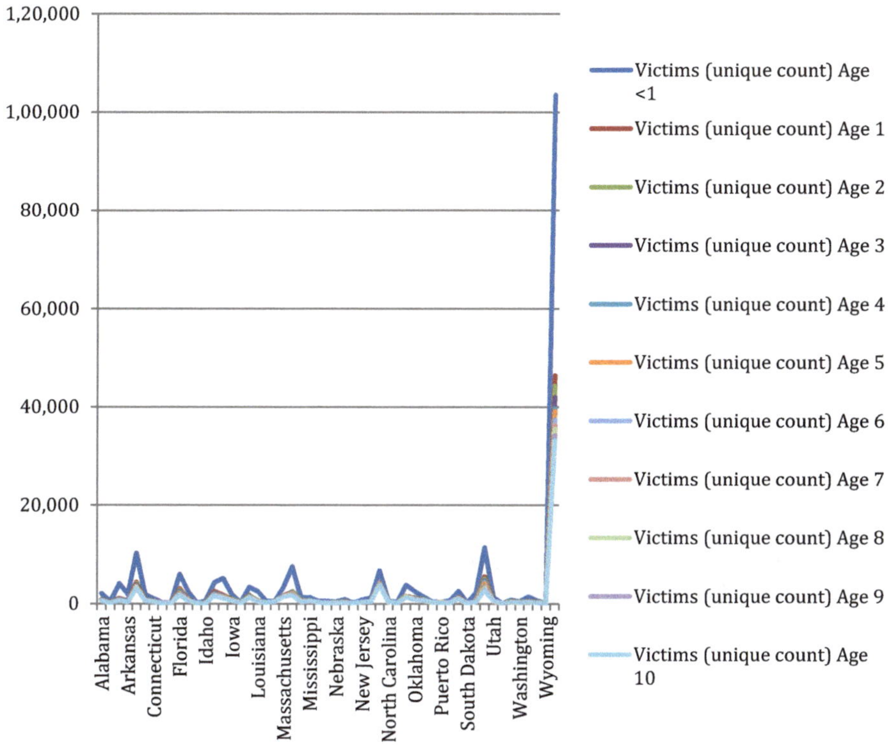

Fig. 4.5 Graph representing Victims of Age (1–10) of different states of the USA

the ranking of the states where there are high number of childhood trauma victims which mainly lies between age 12–16.

4.7.2 Predictive Analytics

ML facilitates the early identification of individuals at risk of PTSD and predicts the potential success of therapeutic interventions.

4.7.3 Enhanced Therapeutic Techniques

ML optimizes therapeutic techniques through real-time feedback analysis, adapting therapy sessions based on immediate patient reactions [23].

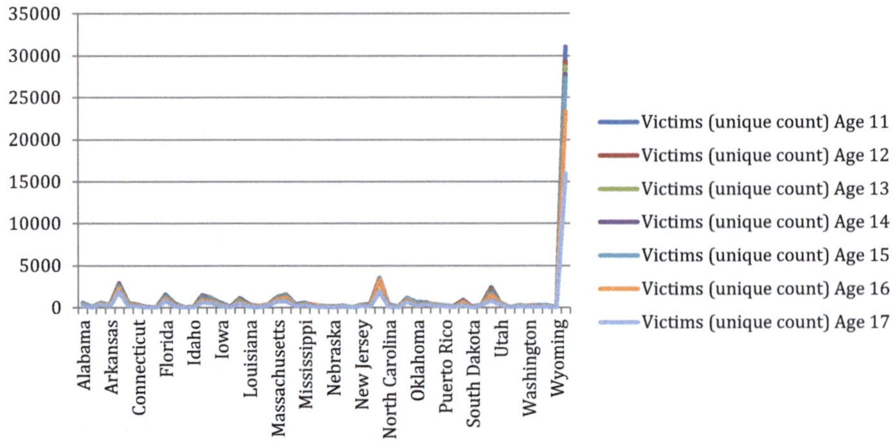

Fig. 4.6 Graph representing Victims of Age (11–17) of different states of the USA

4.7.4 Metrics for Comparing the Effectiveness of Treatments for Childhood Trauma

A range of metrics can evaluate the effectiveness of treatments for childhood trauma, including those addressing PTSD. These metrics enable researchers and clinicians to compare traditional therapeutic approaches with treatments enhanced by machine learning (ML) techniques. Psychophysiological metrics offer an objective way to track treatment progress and recovery in PTSD patients, complementing traditional psychometric assessments often influenced by memory, emotional, or cultural biases. Longitudinal studies reveal that significant symptom improvements correlate with physiological changes, including decreased resting heart rate (HR), reduced heart rate reactivity, enhanced heart rate variability (HRV), lower skin conductance (SC), and adaptive changes in electromyography (EMG) and blood pressure (BP). These markers reflect improved autonomic nervous system (ANS) balance and emotional regulation, supporting their integration into treatment strategies alongside techniques like NLP, V/K Dissociation, and Machine Learning.

Key Metrics:

Heart Rate (HR): Elevated at rest, decreases post-treatment.

Heart Rate Variability (HRV): Improves with better emotion regulation.

Skin Conductance (SC): Reflects arousal; reduces after an effective therapy.

Electromyography (EMG): Measures startle responses; adaptive decreases post-treatment.

Blood Pressure (BP): Both raw values and reactivity lower with recovery.

Psychometric Reports: Subjective assessments to track symptom changes but prone to biases.

This section outlines key categories of metrics used to assess treatment outcomes and their relevance to approaches such as Neuro-Linguistic Programming (NLP), Visual/Kinesthetic Dissociation (V/KD), and ML-based methodologies [19].

4.8 Sensitivity of Childhood Trauma: Ethical Concerns, Patient Consent, and Data Privacy in ML and VR Integration

Childhood trauma is an exceptionally sensitive area in mental health research. The treatment with integrating advanced technologies ML and VR is also sensitive for children cases. These technologies offer the potential to revolutionize therapeutic approaches for PTSD, OCD and childhood trauma. They also raise significant ethical concerns, particularly regarding patient consent and data privacy. Table 4.2 has a discussion of the key considerations in this domain [24, 25].

Integrating ML and VR into therapies for childhood trauma offers transformative potential. It requires navigating significant ethical concerns. Ensuring patient

Table 4.2 The key considerations in sensitivity of childhood trauma: Ethical concerns, patient consent, and data privacy [24]

S. no	Concern	Impact	Mitigation
1	Ethical	Children who have experienced trauma are highly vulnerable. They may have limited capacity to understand or consent to experimental treatments. Re-traumatization and exploitation need to avoid	ML systems rely on data to train predictive models and personalize treatments Bias in ML algorithms could disproportionately impact children from certain demographics Misinterpretations of data patterns could result in inappropriate or ineffective interventions
2	Patient consent	Parents or guardians must provide consent. It is also essential to involve the child in an age-appropriate discussion about the treatment. Legal consent, parental consent and child consent are necessary	Parents and children must clearly understand how their data will be used, including for training ML algorithms or refining VR-based treatments Patients should have the ability to withdraw consent at any stage without consequences for their care
3	Data privacy	The collection and analysis of sensitive personal and medical data raise significant privacy concerns regarding **anonymization and secure storage**	Collaborations between researchers, healthcare providers and technology developers often involve data sharing with limited access and non-commercial usage. Data must be stored on secure platforms compliant with regulations like HIPAA to prevent breaches

consent, safeguarding data privacy and adhering to rigorous ethical standards are imperative to harness these technologies responsibly. Therefore, this innovative approach can provide effective, personalized and safe treatments for childhood PTSD, OCD and trauma by prioritizing the well-being of vulnerable patients.

4.9 Challenges in Implementing ML and VR-Based Approaches in Clinical Settings

The integration of ML and VR into clinical treatments for PTSD, OCD and childhood trauma offers significant promise. But there are several challenges which must be addressed to ensure effective implementation. These limitations lie in financial, technical, ethical and operational domains [26]. There are some hardware and software expenses challenges. Such as VR headsets, high-performance computing systems and specialized ML software can be prohibitively expensive for many clinics. Customizing ML algorithms and VR scenarios for specific therapeutic needs requires significant investment in research and development. Regular updates, troubleshooting, maintenance cost and hardware replacement add to long-term expenses. Since continuous technical support is required for system installation, calibration and troubleshooting. Many ML applications and VR systems rely on fast and stable internet connections. That may not be available in all regions. VR-based therapy set up often requires dedicated physical spaces for equipment and patient interaction. The collection and analysis of sensitive patient data by ML systems raise concerns about data security and compliance with regulations such as HIPAA [27]. ML models may inadvertently perpetuate biases if trained on non-representative datasets, leading to unequal treatment outcomes. Patients may not fully understand the implications of using ML and VR in their treatment, raising ethical concerns. Many ML and VR applications are still in experimental stages and lack large-scale validation studies. To mitigate these challenges some initiatives such as government funding, public–private partnerships and open-source software can help reduce costs. Proper trainings, workshops, certifications, data privacy, broadband expansion, large scale study and R&D are suggestions to enhance applications of ML and VR approaches in clinical settings [28].

4.10 Conclusion

The integration of Neuro-Linguistic Programming, Visual/Kinesthetic Dissociation, and Machine Learning represents a paradigm shift in treating childhood PTSD and trauma. This approach enhances therapeutic efficacy, supports emotional healing, and empowers patients to reclaim their mental health and well-being. Personalized treatment plans, predictive analytics, and adaptive therapies set a new standard for trauma-informed care, providing more effective and compassionate mental health interventions for childhood PTSD and trauma survivors. Results showed that schema therapy can lead to beneficial effects in disorder-specific symptoms and early maladaptive schemas. Yet, we also uncovered substantial methodological limitations in most studies like high initial cost due to requirement of specialized equipment, high performance GPUs, VR headsets and software licences; also implementing these systems required skilled and trained clinical staff. Along with these limitations there are certain ethical concerns related to the use of machine learning and virtual reality in treating childhood trauma are privacy and data security, informed consents by parents, bias and fairness in machine learning, emotional and psychological risks, dependency and technology. Despite growing clinical interest and application, there is a systematic issue with the quality of research in this field. Consequently, we concluded by presenting a research agenda with recommendations for future studies, which are crucial for establishing a solid evidence base for using NLP therapy in treating chronic anxiety, OCD, and PTSD.

References

1. J. M. Carneiro and R. Bertelli (2021), "Interventions Used in the Psychotherapeutic Treatment of Post-Traumatic Stress Disorder (PTSD)," *ASSRJ*, vol. 8, no. 9, pp. 370–405, https://doi.org/10.14738/assrj.89.10924.
2. Richard M. Gray (2010), " NLP and PTSD: The Visual-Kinesthetic Dissociation Protocol", Reprinted from Suppose, the Official CANLP/ACPNL Bilingual Newsletter, pp. 25–42.
3. Figley, C. R., &Ludick, M. (2017)." Secondary traumatization and compassion fatigue. In S. N. Gold (Ed.):, *APA handbook of trauma psychology: Foundations in knowledge,* American Psychological Association, (pp. 573–593), https://doi.org/10.1037/0000019-029.
4. D. B. Samuel, T. Suzuki, M. A Bucher, and S. A. Griffin, (2018), "The agreement between clients' and their therapists' ratings of personality disorder traits.," *Journal of Consulting and Clinical Psychology*, vol. 86, no. 6, pp. 546–555, https://doi.org/10.1037/ccp0000304.
5. Rajeswari. H, B. Sreelekha (2017), "Effectiveness of Neuro Linguistic programming (NLP) on Secondary traumatic stress (STS) among nurses", Global Journal of Research Analysis, vol-6, issue 6, ISSN No. 2277 - 8160
6. N. Peeters, B. Van Passel, and J. Krans (2022), "The effectiveness of schema therapy for patients with anxiety disorders, OCD, or PTSD: A systematic review and research agenda," *British J Clinic Psychol*, vol. 61, no. 3, pp. 579–597, https://doi.org/10.1111/bjc.12324.
7. Grinder, J., & Bandler, R. (1975). The Structure of Magic I: A Book About Languageand Therapy. Palo Alto, CA: Science and Behavior Books.
8. Dimmick, S., (2017). Successful Communication Through NLP: A Trainer 's Guide. Routledge.
9. Juhnke, G. A., Coll, K. M., Sunich, M. F., & Kent, R. R. (2008). Using a Modified Neurolinguistic Programming Swish Pattern With Couple Parasuicide and Suicide Survivors. *The Family Journal*, *16*(4), 391–396. https://doi.org/10.1177/1066480708322807.
10. E. Vanmassenhove, J. P. Cabral, and F. Haider (2016),"Prediction of Emotions from Text using Sentiment Analysis for Expressive Speech Synthesis," in *9th ISCA Workshop on Speech Synthesis Workshop (SSW 9)*, ISCA, pp. 21–26. https://doi.org/10.21437/SSW.2016-4.

11. Drigas, A., Mitsea, E., & Skianis, C. (2022). Neuro-linguistic programming, positive psychology & VR in special education. *Scientific Electronic Archives, 15*(1).
12. S. Sahu, S. K. Rout, and D. Mohanty (2015), "Twitter Sentiment Analysis -- A More Enhanced Way of Classification and Scoring," in *2015 IEEE International Symposium on Nanoelectronic and Information Systems*, Indore, India: IEEE, pp. 67–72. https://doi.org/10.1109/iNIS.2015.40.
13. K. Dormandy and B. Grimley (2024), "Gatekeeping in Science: Lessons from the Case of Psychology and Neuro-Linguistic Programming," *Social Epistemology*, vol. 38, no. 3, pp. 392–412, https://doi.org/10.1080/02691728.2024.2326828.
14. Hickey, K.L., (2024). Comparison of language and somatic experiences between reports of trauma and trauma-related dreams & personality features of trauma-exposed persons reporting trauma-related dreams (Doctoral dissertation).
15. Jahanzeb Jahan, Minahil Tariq, Mubashar Nadeem (2022), The Effects of Neuro-linguistic Programming on a Psychotherapist's Communication Patterns: A Case Study, JDSS 3, https://doi.org/10.47205/jdss.2022(3-II)13.
16. Przemysław Turkowski, Jan Jędrzejczyk, Mirosława Huflejt-Łukasik, Joanna Wieliczko (2016), Neuro-linguistic psychotherapy in the treatment of anxiety disorders, vol. 178.
17. Grimley, B. N. (2021). Sailing the 7Cs of Courage. A Framework for One-to-one Work Emerging from Neuro-Linguistic Programming (NLP)-a Single Case History. *Journal of Experiential Psychotherapy/Revista de PSIHOterapieExperientiala, 24*(1).
18. El-Ashry, M. M., (2021), The importance of Neuro-Linguistic Programming skills as a communication tool in the workplace. Journal of Global Scientific Research, 6(1), pp.1108–1123.
19. R. C. Freire, C. Cabrera-Abreu, and R. Milev (2020), "Neurostimulation in Anxiety Disorders, Post-traumatic Stress Disorder, and Obsessive-Compulsive Disorder," in *Anxiety Disorders*, vol. 1191, Y.-K. Kim, Ed., in Advances in Experimental Medicine and Biology, vol. 1191., Springer Singapore, pp. 331–346. https://doi.org/10.1007/978-981-32-9705-0_18.
20. P. Cheng *et al.* (2020),"Psychometric property study of the posttraumatic stress disorder (PTSD) checklist for DSM-5 (PCL-5) in Chinese healthcare workers during the outbreak of coronavirus disease 2019," *Journal of Affective Disorders*, vol. 277, pp. 368–374, https://doi.org/10.1016/j.jad.2020.08.038.
21. D. D. Mendes, M. F. Mello, P. Ventura, C. De Medeiros Passarela, and J. De Jesus Mari (2008), "A Systematic Review on the Effectiveness of Cognitive Behavioral Therapy for Posttraumatic Stress Disorder," *Int J Psychiatry Med*, vol. 38, no. 3, pp. 241–259, https://doi.org/10.2190/PM.38.3.b.
22. U.S. Department of Health & Human Services (2018), National Child Abuse and Neglect Data System (NCANDS) Child File: Findings from the National Child Abuse and Neglect Data System (NCANDS) - Data tables (from Child Maltreatment 2018), https://catalog.data.gov/dataset/national-child-abuse-and-neglect-data-system-ncands-child-file-findings-from-the-national--4dd6a.
23. A. D'Andrea, F. Ferri, P. Grifoni, and T. Guzzo (2015), Approaches, Tools and Applications for Sentiment Analysis Implementation, *IJCA*, vol. 125, no. 3, pp. 26–33, https://doi.org/10.5120/ijca2015905866.
24. Adebayo O, Bhuiyan ZA, Ahmed Z (2023). Exploring the effectiveness of artificial intelligence, machine learning and deep learning in trauma triage: A systematic review and meta-analysis. Digit Health. Oct 9;9:20552076231205736. https://doi.org/10.1177/20552076231205736. PMID: 37822960; PMCID: PMC10563501.
25. David B. Olawade, Ojima Z. Wada, AderonkeOdetayo, Aanuoluwapo Clement David-Olawade, FiyinfoluwaAsaolu, Judith Eberhardt (2024). Enhancing mental health with Artificial Intelligence: Current trends and future prospects, Journal of Medicine, Surgery, and Public Health, 3, 100099, ISSN 2949–916X, https://doi.org/10.1016/j.glmedi.2024.100099
26. Betelhem Zewdu Wubineh, Fitsum Gizachew Deriba, Michael Melese Woldeyohannis (2024), "Exploring the opportunities and challenges of implementing artificial intelligence in healthcare: A systematic literature review", Urologic Oncology: Seminars and Original Investigations, Volume 42, Issue 3, 48–56, ISSN 1078–1439, https://doi.org/10.1016/j.urolonc.2023.11.019

27. Kang CY, Yoon JH. Current challenges in adopting machine learning to critical care and emergency medicine. ClinExpEmerg Med. (2023) Jun;10(2):132–137. https://doi.org/10.15441/ceem.23.041. Epub 2023 May 15. PMID: 37188356; PMCID: PMC10350350.
28. Yoon, J. H., Pinsky, M. R., & Clermont, G. (2022). Artificial intelligence in critical care medicine. *Annual Update in Intensive Care and Emergency Medicine 2022*, 353–367.
29. Bruce Grimley (2016), Removal of PTSD Symptoms in a Client Using Neuro Linguistic Programming, Journal of Experiential Psychotherapy, vol. 19, no 3 (75).
30. Michelle Yang, Noor Mady, Outi Linnaranta, Utility of Psychophysiological Metrics in Guiding Treatment of Trauma Symptoms: A Systematic Review, Journal of Behavioral and Cognitive Therapy, Vol.31, Issue 3, 2021, Pp. 249–266, ISSN 2589–9791, https://doi.org/10.1016/j.jbct.2021.01.004.

Chapter 5
"Robotics, Automation, and Telemedicine in Healthcare 5.0: Diagnostic and Treatment Processes"

Reeta Clonia and Ritu Vats

Abstract The advancement of robots and automation in the medical field has completely changed the way that diagnostics and treatments are carried out, greatly enhancing accuracy, effectiveness, and patient outcomes. This study examines the application of robots to minimally invasive surgery, personalised medicine, rehabilitation, and diagnostic imaging. It talks about the clinical advantages of these technologies as well as the difficulties healthcare systems have in using them. The future of medicine depends on our ability to comprehend the uses and consequences of automation, which is becoming increasingly important.

Keywords Medical robotics · Automated diagnosis · Robot-assisted surgery · AI in healthcare · Telemedicine and robotics · Surgical robotics

5.1 Introduction

The use of automation and robots in healthcare has revolutionised the field, particularly in terms of diagnosis and treatment. Robotics, which was first applied in sectors such as manufacturing, has now entered the medical field and is offering previously unheard-of levels of precision and operational efficiency. Robotic systems have greatly improved medical professionals' abilities by providing instruments that are more accurate and dexterous than human hands, which has improved patient outcomes. Automation has made it possible to diagnose conditions more quickly, eliminate human error, and provide individualised treatment regimens. This is especially true when automation is combined with artificial intelligence (AI)

R. Clonia (✉) · R. Vats
Maharishi Markandeshwar Institute of Management, Maharishi Markandeshwar (Deemed to Be University), Mullana, Ambala, Haryana, India
e-mail: reeta.clonia@gmail.com

R. Vats
e-mail: sritu894@gmail.com

© The Author(s), under exclusive license to Springer Nature Switzerland AG 2026
K. Dalip et al. (eds.), *Healthcare 5.0*, https://doi.org/10.1007/978-3-032-09582-4_5

and machine learning (ML). These technologies have the potential to completely transform healthcare as they advance.

The necessity for accuracy and consistency in medical procedures—particularly in surgery and diagnostics—is a major driving force for the incorporation of robots in healthcare. Automated solutions can assist decrease the variability caused by human variables, ensuring that operations are done with equal precision. Furthermore, robotic surgical technologies that allow for more accurate interventions with quicker recovery periods have been made possible by the increased desire for minimally invasive therapies. In this thorough analysis, we examine the many ways that robots and automation are used in healthcare diagnoses and treatments, as well as their present capabilities, difficulties, and potential future developments.

5.2 Robotics in Diagnostics

5.2.1 *Advanced Imaging Techniques*

One of the most essential parts of contemporary diagnosis is medical imaging. Increased accuracy and detail in human body visualisations can be attributed to the significant advancements in imaging techniques brought about by robotics and automation. In order to provide the best alignment for imaging, robotic devices in MRI and CT scanners, for example, may autonomously position patients. Repetitive scans are not as necessary because to this automation, which also lowers the possibility of mistakes. In addition, imaging systems are increasingly using AI-powered algorithms to identify problems that human eyes might overlook, improving the accuracy of diagnoses [1].

Another significant development in diagnostic imaging are automated ultrasound devices. In the past, ultrasonic scans needed the transducer to be manually moved, which might vary the image quality depending on the operator's skill level. Thanks to robotic ultrasound devices, scanning may now be done automatically, producing reliable, high-quality pictures and lowering the need for operators [2]. These devices can be especially helpful for foetal imaging, where precise diagnosis and monitoring depend on constant picture quality.

5.2.2 *Robotic-Assisted Biopsies*

Robotically assisted biopsy systems are increasing the accuracy of diagnosis, especially in the field of cancer diagnosis. These tools significantly improve the precision of biopsies by accurately guiding needles to the location of a cancer. To reduce the need for random biopsies and enable more effective treatment planning, robotic guidance systems, for example, have made it simpler to target specific prostate areas during the diagnosis of prostate cancer [3]. The precision of robotics lessens the chance of missing small or difficult-to-reach lesions.

Studies have shown that robotically assisted biopsies reduce issues and speed up patients' recovery. This is particularly important for delicate procedures where precision is necessary to avoid tissue injury, such brain biopsies. Furthermore, biopsies are performed with the highest degree of accuracy possible because to the integration of robotics and real-time imaging technologies like MRI [4].

5.2.3 Capsule Endoscopy and Robotic Endoscopes

By providing more accurate and manoeuvrable instruments for examining the inside structures of the body, robotic endoscopes are revolutionising the area of internal diagnostics. With capsule endoscopy, doctors may take precise pictures of the gastrointestinal (GI) tract without doing intrusive operations because it uses a tiny camera housed in a robot the size of a pill. Sending high-definition photos to a receiver outside the body, the capsule navigates the digestive system on its own [5].

Apart from capsule endoscopy, robotic endoscopes are employed in bronchoscopy and colonoscopy operations, since they offer improved accuracy and control. Conventional endoscopes are operated by hand, which presents difficulties when negotiating intricate anatomical structures. However, more precise movement and positioning are made possible by robotic devices, which enhance patient comfort and diagnostic capabilities [14].

5.3 Robotics in Treatment Processes

5.3.1 Minimally Invasive Surgery

One of the most well-known uses of robotics in medicine is robotic-assisted surgery. Minimally invasive surgery has been transformed by systems like the da Vinci Surgical System, which provide better control, accuracy, and flexibility than conventional laparoscopic procedures. Using a console, surgeons may control the robotic arms, giving them more dexterity and 3D visualization—both essential for carrying out complex surgeries.

Robotic surgery's accuracy is especially useful for delicate procedures like those on the heart, brain, or reproductive organs. For example, robotic-assisted prostatectomies in urology have demonstrated better results than conventional techniques, including shorter recovery periods and less problems [6]. Similarly, hysterectomies and other gynaecological procedures are frequently performed with robotic systems, which enable smaller incisions, less blood loss, and faster patient recovery [7].

Additionally, robotics increases surgical resection accuracy, especially in cancer procedures. Surgeons may more accurately target malignant areas while sparing healthy tissue by using robotic devices with real-time imaging technology, which improves patient outcomes overall [8].

5.3.2 Orthopedic Robotics

Robotic help has proven invaluable in orthopaedic procedures, especially joint replacements. Surgeons may now design and perform joint replacements with an accuracy previously impossible because to robotic technologies like the MAKO robotic-arm-assisted technology. The robotic system builds a three-dimensional (3D) model of the patient's anatomy using preoperative imaging data, which enables the surgeon to design the best possible implant placement.

For the joint replacement to last a long time, the robotic arm helps the physician during surgery by making sure the bone is sliced precisely and the implant is positioned appropriately [9]. Because of its accuracy, there is a lower chance of issues such implant misalignment, which can cause early failure and necessitate revision surgery. Research has indicated that patients who have robotic-assisted knee and hip replacements get improved functional results and quicker recovery periods [10].

5.3.3 Automation in Radiation Therapy

Automation has greatly increased the efficacy of radiation therapy, which is a mainstay in the treatment of many malignancies. Radiation therapy has benefited from automated technologies, such as robotic arm-equipped linear accelerators, which have increased radiation delivery precision and reduced exposure to nearby healthy tissues [11]. These devices make sure that the tumour is precisely targeted by using artificial intelligence (AI) algorithms to determine the best radiation dose and direction.

In addition to increasing the distribution of radiation, automation has improved the planning process. The time needed for treatment planning may be decreased, and overall process efficiency can be increased, by using automated systems to analyse diagnostic pictures and generate highly customised treatment plans [12].

5.3.4 Robotic Drug Delivery

One interesting area of study is the development of robotic devices for targeted medicine delivery. These systems utilise nanorobots or microbots to administer medications specifically to the site of sickness, reducing systemic adverse effects. Robotic drug delivery systems, for instance, can accurately target tumours during cancer treatment, enabling larger dosages of chemotherapeutic medications at the tumour location while protecting healthy tissues [13].

In the treatment of neurodegenerative illnesses, where getting medications beyond the blood–brain barrier has historically proven difficult, robotic drug administration is also being investigated. Crossing this barrier using nanorobots opens up new treatment options for diseases like Parkinson's and Alzheimer's [14].

5.4 Robotics in Rehabilitation

For individuals suffering from musculoskeletal diseases, strokes, or neurological impairments, rehabilitation is an essential part of their healing process. Robotics is increasingly being utilised to better rehabilitation results by offering more precise and controlled motions during treatment.

5.4.1 Robotic Exoskeletons

Wearable technology called robotic exoskeletons helps patients regain movement by giving weak or paralysed limbs external assistance. Patients healing from spinal cord injuries and stroke patients especially benefit from these exoskeletons. For example, patients can practise walking with the help of a robotic framework by using the Lokomat system, a robotic exoskeleton utilised in gait training [15].

Studies have demonstrated that by offering regular and repeated movement—which is critical for neuroplasticity and the restoration of motor function—robotic exoskeletons enhance patient outcomes [16]. These devices also allow for customised rehabilitation regimens based on the unique requirements of each patient.

5.4.2 Robotics in Neurorehabilitation

Robotics has been a huge help to neurorehabilitation, particularly for stroke, traumatic brain injury, and spinal cord injuries. By automating repeated motions and providing patients and therapists with real-time feedback, robotic systems can support physical therapy. By directing stroke patients through repetitive motions that aid in rehabilitation, devices such as the Armeo robotic arm help patients regain function in their upper limbs [17].

Furthermore, to develop immersive rehabilitation experiences, robotics has been combined with virtual reality (VR) settings. In therapeutic sessions, these environments promote motivation and engagement by allowing patients to participate with a virtual world [18]. The integration of robotics and VR has been found to increase patient adherence to rehabilitation regimens and speed recovery timeframes.

5.5 Results and Discussion

5.5.1 Advancements in Robotics and Automation in Healthcare 5.0

Healthcare 5.0 is changing the face of diagnosis and treatment by emphasising automation, accuracy, and personalisation. Advances in robotics and automation have made minimally invasive procedures, precise diagnoses, and quicker recovery periods possible. Modern advancements like robotically assisted operations have improved the precision of intricate procedures and decreased human error. The speed and accuracy of detecting illnesses has increased with automation in diagnostic procedures, such as AI-powered imaging, enabling earlier identification and better treatment outcomes.

Robotic devices, such as telerobotics, allow for real-time patient monitoring and remote procedures in the context of telemedicine, therefore reducing the gap that separates patients and physicians geographically. With the assistance of human surgeons, robots may now carry out operations remotely, expanding access to healthcare, particularly in distant and underdeveloped places.

5.5.2 Case Studies and Comparative Analysis

Case examples abound that show how automation and robots improve healthcare results. By adopting the Da Vinci surgical system, for example, robotically assisted operations have led to a decrease in complications, a reduction in blood loss, and a faster recovery period for the patients. Hospital operations are more efficiently run when patients receiving robotic surgery remain less in the hospital than those using traditional procedures, according to studies.

Similar to this, faster diagnoses have resulted from automation in diagnostic technologies like AI imaging algorithms. AI-driven automation has improved diagnosis accuracy in certain circumstances, such as lung cancer screening, which has led to early treatment interventions and a decrease in the need for repeat testing.

5.5.3 Challenges and Trends in Integrating Telemedicine with Robotics

Even though telemedicine and robots have shown a lot of promise, there are still issues in integrating these technologies into Healthcare 5.0. The growing use of artificial intelligence (AI) and real-time data to improve decision-making is one significant development. AI-powered diagnostic tools on telemedicine networks may autonomously analyse patient data and recommend treatments to doctors, expediting and enhancing accuracy.

Assuring smooth integration between robotic equipment and telemedicine systems is a difficulty, though. It is necessary to handle issues like data security, privacy, and medical device interoperability across many platforms. Furthermore, patients require assistance in order to feel at ease with robotically assisted tests and treatments, and healthcare professionals require training in order to use this cutting-edge technology.

5.5.4 A Prospective View on Patient-Centered Healthcare 5.0 Medical

In 5.0, automated and robotic systems optimise individualised treatment in a more patient-centered manner. Healthcare professionals may obtain detailed patient profiles via telemedicine systems, which when paired with AI and real-time data, allow for more precise diagnosis and customised treatment plans. The transition to automated, data-driven healthcare improves treatment results and patient happiness while guaranteeing that medical procedures are less intrusive, quicker, and more precise.

- **Diagram of Robotics in Surgery (Healthcare 5.0 Context)**: An illustration of the roles that robotic systems, AI, and doctors play in precision surgery, such as a flowchart or diagram, explain how robotics enables surgery.
- **Telemedicine Integration with Robotics**: An example of how telemedicine systems might facilitate communication between medical professionals and patients by using robots and artificial intelligence (AI) capabilities to help with remote diagnoses and procedures. Patient engagement, data flow, and real-time therapy might all be highlighted in the graphic.
- **AI-Powered Diagnostic Automation:** An illustration contrasting AI-driven automation (faster, more accurate outcomes) with conventional diagnostic methods (manual picture inspection, for example). A before-and-after example with healthcare 5.0 advancements may be used to illustrate this.

5.6 Challenges in the Adoption of Robotics and Automation in Healthcare

Even while robots and automation have many advantages in diagnosis and treatment, there are still a number of obstacles in the way of their broad use.

5.6.1 Cost and Accessibility

The high cost of these technologies is one of the main obstacles to the adoption of robotic systems. Many healthcare facilities, especially those in low-resource countries, may not be able to afford the prohibitively expensive advanced robotic surgical equipment and automated diagnostic instruments [19]. One major obstacle is the cost of educating medical staff to utilise these technologies.

5.6.2 Training and Skill Development

Health care workers must receive in-depth training in order to use robotics in the field. It can take a lot of time and money to operate these devices, which requires expertise from radiologists, surgeons, and therapists. Furthermore, learning robotic systems has a high learning curve that may initially impede adoption [6].

5.6.3 Ethical and Privacy Concerns

Concerns about patient privacy and data security are among the ethical issues brought up by the growing use of automation and artificial intelligence in healthcare. Large datasets are frequently necessary for automated systems to operate efficiently, thus it's critical to make sure that this data is managed carefully. Furthermore, concerns regarding accountability and the function of human oversight become crucial as automation assumes greater decision-making responsibilities in the healthcare industry [20].

5.7 Future Trends in Robotics and Automation in Healthcare

Future developments in robotics and automation technology will probably influence the healthcare industry in a number of ways.

5.7.1 AI-Driven Robotic Systems

AI is predicted to play an increasingly crucial role in robotics, particularly in fields like diagnostics and customised medicine. Massive volumes of patient data may be analysed by machine learning algorithms to produce highly customised treatment plans that robotic devices can then precisely carry out [19]. Robotics and AI together promise to improve medical treatments' accuracy and speed.

5.7.2 Nanorobotics

The use of microscopic robots to carry out activities at the cellular level is known as nanorobotics, and it is an emerging topic with great medical promise. These nanorobots may be employed in treatments such as tissue healing, medication distribution, and even diagnostics. One potential use of nanorobots is the delivery of medications directly to cancer cells, therefore reducing tissue damage [13].

5.7.3 Robotic-Assisted Telemedicine

Telemedicine has become more popular as a result of the COVID-19 epidemic, and robotics will probably be used in distant healthcare in the future. Remote surgical and diagnostic operations can be completed by robotic devices, giving patients in remote locations access to top-notch medical treatment. Healthcare professionals may be able to conduct physical examinations remotely thanks to telemedicine robots that are outfitted with diagnostic instruments [21].

5.8 Conclusion

Robotics and automation in diagnostic and therapeutic processes have completely changed the healthcare industry by improving patient outcomes, productivity, and accuracy. By enabling more precise diagnosis, less invasive procedures, and cutting-edge treatment alternatives, these technology advancements are completely changing the healthcare industry. While robot-assisted operations reduce human error and improve precision in sensitive procedures, robotic rehabilitation is providing patients with more effective and customised recovery strategies. In addition to providing real-time insights and enhancing the speed and accuracy of diagnostic evaluations, automated drug administration systems and robotic imaging technologies help to streamline healthcare procedures.

With the use of AI and machine learning in diagnostics, large datasets may be analysed to find trends and anticipate possible health problems ahead of time. This can improve preventative care and expedite therapeutic procedures. These tools assist medical practitioners in making better judgements by increasing diagnostic accuracy. AI, for instance, may help with picture analysis for medical purposes, anomaly detection, and therapy recommendation.

But in order to guarantee that these developments help every patient, a number of issues need to be resolved. The expense of integrating robots and automation in healthcare is a significant worry as it may restrict access, particularly in impoverished or rural healthcare systems. Another challenge is that operating these sophisticated devices needs specialised training for healthcare personnel, which takes a lot of time and money. It's also important to properly address ethical issues, such as patient data privacy and the risk of relying too much on automated systems.

Healthcare robots have a bright future despite these obstacles. Continuous developments in AI, machine learning, and nanotechnology have the potential to substantially enhance patient care by increasing the accuracy of diagnoses and personalising therapies. Automation and robots will continue to be crucial in determining the direction of healthcare as long as the proper plans are put in place to get past the present obstacles.

References

1. Kowal, M. M., Singh, A., & Lee, J. K. (2021). Robotic imaging systems: Revolutionizing diagnostic accuracy. *Journal of Medical Robotics*, 13(4), 98-108.
2. Smith, A. D., & Murray, E. A. (2022). Automated ultrasound: Advancing diagnostic imaging. *Radiology and Imaging Journal*, 24(1), 34-46.
3. Jones, M. A., Davis, L. H., & Wilson, A. (2022). Robotic-guided biopsies in prostate cancer: A clinical review. *Journal of Urology and Oncology*, 12(3), 112-120.
4. Garcia, J. A., & Patel, V. (2021). Robotic-assisted biopsies: Improving precision in cancer diagnosis. *International Journal of Cancer Diagnostics*, 25(6), 321-334.
5. Sato, M., & Takahashi, K. (2021). Capsule endoscopy: A robotic revolution in GI diagnostics. *Gastroenterology Review*, 13(1), 54-66.
6. Simpson, A. K., Chen, Y. J., & Lister, S. (2023). Robotic surgery in urology: Advances and challenges. *Journal of Robotic Urology*, 19(3), 67-81.
7. Wilson, K. L., Davis, J., & Hall, P. A. (2021). Robotic gynecological surgery: Reducing complications and improving outcomes. *Gynecological Surgery Review*, 27(4), 89-104.
8. Huang, Y., Yang, X., & Zhao, J. (2023). Robotic-assisted surgery: A review of clinical benefits and future directions. *Surgical Innovations Journal*, 30(2), 81-95.
9. Krause, M., Patel, V., & McKnight, C. (2022). Robotic-assisted joint replacement: A new standard of care in orthopedics. *Orthopedic Technology Journal*, 9(2), 211-219.
10. Martin, B. R., Davis, S. A., & Jensen, K. M. (2022). Robotic-assisted surgery in orthopedics: A systematic review. *Orthopedics International*, 26(3), 213-230.
11. Brenner, D. J., & Hall, E. J. (2021). Radiation and the precision revolution in oncology. *Journal of Radiation Oncology*, 15(2), 204-217.
12. Gupta, P., Verma, S., & Martin, H. (2023). Automation in radiation oncology: Integrating AI and robotics for precision. *Journal of Cancer Treatment*, 17(1), 14-22.
13. Patel, P., & Li, Z. (2021). Nanorobotics in targeted drug delivery: Revolutionizing cancer therapy. *Journal of Nanotechnology and Medicine*, 5(2), 93-101.

14. Kim, S. K., & Park, H. J. (2022). Nanorobotics in the treatment of neurodegenerative diseases. *Journal of Nanomedicine and Therapeutics*, 9(1), 32-47.
15. Reinkensmeyer, D. J., & Emken, J. L. (2022). Robotic exoskeletons in stroke rehabilitation: Evidence and future directions. *Stroke Recovery Journal*, 18(2), 145-162.
16. Zhang, J., Wang, S., & Liu, C. (2021). The future of robotic exoskeletons in neurorehabilitation. *Journal of Neuroengineering and Rehabilitation*, 18(2), 203-215.
17. Krebs, H. I., Volpe, B. T., & Hogan, N. (2022). Robotic neurorehabilitation: From research to practice. *Journal of Rehabilitation Robotics*, 22(1), 78-95.
18. Brochard, S., Roby-Brami, A., &Rémy-Néris, O. (2022). Robotics and virtual reality in rehabilitation: Improving patient engagement. *Neurorehabilitation Review*, 14(3), 145-160.
19. Chen, J., Davis, S. R., & Zhang, H. (2023). Robotics in healthcare: Current trends and future opportunities. *Journal of Robotics in Medicine*, 12(1), 56-78.
20. Leite, L., Sousa, P., & Antunes, M. (2021). Ethical considerations in the use of robotics and AI in healthcare. *Healthcare Ethics Quarterly*, 5(2), 45-57.
21. Fitzpatrick, R., Levin, S., & Tam, L. (2022). Robotic-assisted telemedicine: The future of remote healthcare. *Telemedicine Journal*, 9(4), 78-89.
22. Hussain, M., Zubair, M., & Sheikh, A. (2021). Robotics in healthcare: A review. *Journal of Healthcare Engineering*, 2021. https://doi.org/10.1155/2021/9831826
23. Zhao, X., Liu, X., & Yang, Q. (2020). The impact of telemedicine on healthcare delivery: A comprehensive review. *Telemedicine Journal and e-Health*, 26(5), 626-632. https://doi.org/10.1089/tmj.2019.0278
24. Moustaide, M., El Afia, A., &BelkahlaDriss, O. (2021). The role of AI and automation in enhancing telemedicine services. *Artificial Intelligence in Medicine*, 115. https://doi.org/10.1016/j.artmed.2021.102057
25. Panch, T., Mattie, H., & Atun, R. (2019). Artificial intelligence and algorithmic bias: Implications for health systems. *Journal of Global Health*, 9(2), 1-6. https://doi.org/10.7189/jogh.09.020318
26. Herrmann, J., Clancy, K., & Henneman, D. (2020). Advancing healthcare 5.0 with robotics and automation in surgery. *Journal of Surgical Research*, 255, 243-252. https://doi.org/10.1016/j.jss.2020.05.036
27. Vimal, S., Saravanan, M., & Natesan, S. (2022). Healthcare 5.0: Enabling technologies for digital transformation in healthcare. *Healthcare Technology Letters*, 9(4), 143–151. https://doi.org/10.1049/htl.2022.0038

Chapter 6
The Role of Artificial Intelligence in Legal Decision-Making

Mirza Samiulla Beg

Abstract Artificial Intelligence (AI) is changing the legal system quickly and dramatically by providing new tools that have a big influence on how decisions are made in the legal system. This chapter explores the diverse uses, advantages, and related difficulties of artificial intelligence in the legal field. Predictive analytics, which uses historical data to predict case outcomes, legal research and document review, which use sophisticated natural language processing to speed up information retrieval, and automated contract analysis, which improves accuracy and efficiency in contract management, are some of the key applications of artificial intelligence. These developments lead to increased effectiveness and better access to the legal system, especially for marginalised groups. But the use of AI also brings up important questions about justice, prejudice, and openness. Because AI systems are taught on past data, they could unintentionally reinforce pre-existing prejudices, therefore close supervision is necessary to guarantee fair results. Furthermore, it's critical to combine human judgement with AI-driven insights and take ethical issues like data security and privacy into account. In order to solve these issues and fully utilise AI's potential to uphold rather than undermine the concepts of justice, this chapter emphasises the necessity of continued collaboration between legal experts, engineers, and politicians. This chapter offers an extensive assessment of the existing situation and potential paths for incorporating AI into the legal system by examining both the revolutionary advantages and the inherent hazards of using AI in legal decision-making.

6.1 Introduction

The judicial system in particular is being progressively impacted by artificial intelligence (AI) in many facets of contemporary society. Artificial Intelligence (AI) tools are revolutionising legal decision-making, ranging from automated contract analysis to predictive analytics. This chapter explores the many uses of AI in the

M. S. Beg (✉)
Department of Computer Science and Engineering, Poornima University, Jaipur, India
e-mail: msaksuniversity@gmail.com

© The Author(s), under exclusive license to Springer Nature Switzerland AG 2026
K. Dalip et al. (eds.), *Healthcare 5.0*, https://doi.org/10.1007/978-3-032-09582-4_6

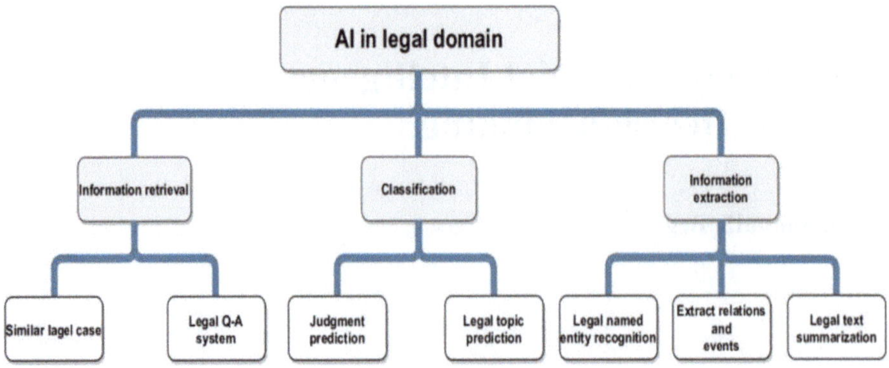

Fig. 6.1 Roles of artificial intelligence in legal system

legal field, evaluates how they affect the legal system, and talks about the moral and legal issues that come with using them. Artificial intelligence (AI) has revolutionised several sectors, radically changing how activities are carried out and judgements are made. The legal industry has seen a dramatic transformation in recent years as AI technologies start to transform established procedures.

The integration of artificial intelligence (AI) into the legal system is a significant advancement that will improve the accessibility, accuracy, and efficiency of the legal system. Artificial intellect (AI) includes a variety of technologies, such as robots, machine learning, and natural language processing (NLP), that are intended to mimic human intellect [1]. These technologies allow systems to automate difficult activities, comprehend natural language, and learn from data. The legal industry is changing as a result of the use of AI, which takes advantage of these capabilities to improve decision-making, give deeper insights, and expedite legal procedures [2]. The different types of roles of artificial intelligence as per the legal system is explained in the Fig. 6.1.

6.2 Historical Context and Evolution

It is crucial to comprehend AI's historical evolution in order to fully appreciate its current position in law. The introduction of computerised databases and case management systems marked the beginning of the integration of technology into legal processes. These early technologies were intended to streamline legal information organisation and automate repetitive activities. But the real breakthrough came with the introduction of more advanced AI technologies like machine learning and natural language processing. Expert systems were introduced for use in legal applications in the 1970s and 1980s [3]. These systems provide the foundation for more sophisticated AI technologies by being able to mimic human experts' decision-making processes.

AI started to provide more dynamic and data-driven insights in the 1990s and 2000s with the introduction of machine learning.

Legal writings and documents may now be handled by AI systems more skilfully thanks to the advancement of natural language processing (NLP). These days, artificial intelligence (AI) in law covers a wide spectrum of uses, from sophisticated case management systems and virtual legal assistants to automated document analysis and predictive analytics. These developments underscore the potential of AI to completely transform conventional procedures and represent a larger trend towards its integration across a range of professional fields like taxation, patent, RTI etcetera [4].

6.2.1 Early Developments in Legal Technology

Basic electronic tools and databases were the first examples of how technology was incorporated into the judicial system. The main goals of early legal technology were case file organisation and administrative work optimisation. AI's introduction brought with it more advanced tools that could do intricate analysis and forecasts. Legal Technology's Early Years When fundamental legal technology first appeared in the middle of the 20th century, the path towards AI in law was set in motion [5]. Legal research databases and computerised case management systems were examples of early legal technology. The purpose of these systems was to replace manual processes with automated ones, such filing and keeping case records. Expert systems first appeared as a type of artificial intelligence in the 1970s and 1980s. Expert systems are computer programs designed to simulate human experts' decision-making processes. These systems were applied to the legal domain to create tools like as the MYCIN system, which was initially intended for medical diagnosis but was later utilised as a template for legal expert systems. These early expert systems relied on established rules and knowledge sources to provide legal counsel and assistance. In this context.

6.2.2 Evolution to Advanced AI

With the development of machine learning and natural language processing in the twenty- first century, artificial intelligence is now capable of more than just automation. These days, AI systems might provide insights, evaluate enormous volumes of legal data, and even help with documentdraughting.

A major advancement in legal technology is the application of artificial intelligence (AI) in the legal sector. Examining the historical evolution of AI in law—including its early beginnings, significant developments, and how these discoveries have affected modern legal practices—is essential to understanding its present position and effect [5]. The working model of legal system is shown in the Fig. 6.2. It state that there are four main paradigm of legal system procedure that is Input,

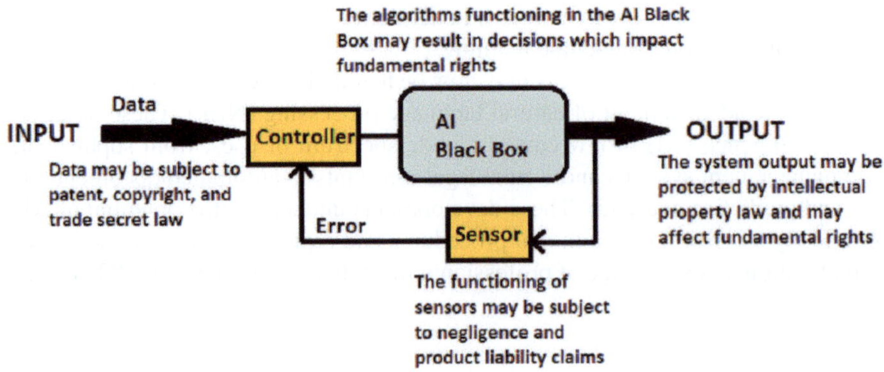

Fig. 6.2 Working Model of legal procedure with AI

controller, AI Black box, sensor and output. The raw input will be passed through the input units, Here the data may be subject to patent, copyrights, and trade secret law. Here the working of the sensor is subject to the negligence and product liability claims. The AI Black box will process and analyse the processing of the artificial intelligence algorithm where the main result will be calculated and compiled which directly or indirectly will impact the fundamental rights, patentability, laws, regulation and various other criteria's. The next unit is output from where the output will be created and displayed.

6.3 The Rise of AI and Machine Learning

6.3.1 The Advent of Machine Learning

Machine learning, a branch of artificial intelligence that involves teaching algorithms to identify patterns and forecast outcomes based on data, experienced tremendous breakthroughs in the 1990s and 2000s. Around this time, machine learning models were able to analyse massive datasets and find patterns and insights, which paved the way for the emergence of more advanced AI applications in the legal field.

6.3.2 Early AI Applications in Legal Research

AI applications in legal studies started to take shape at this time. Basic keyword search features in tools like Westlaw and LexisNexis gave way to more sophisticated ones. These systems began using AI methods to improve search relevancy and accuracy, providing more complex and context-aware results.

6.3.3 Natural Language Processing (NLP)

Significant progress was made in Natural Language Processing (NLP) in the 2010s. NLP is a subfield of artificial intelligence that studies how computers and human language interact. AI systems are now better able to comprehend, interpret, and produce human language thanks to NLP technology [6]. This resulted in the creation of instruments for the legal sector that could more precisely carry out activities including contract analysis, case law research, and legal document assessment.

6.3.4 Predictive Analytics and Data-Driven Insights

Predictive analytics has emerged as a major use of AI in the legal field, tilizing past data to predict case outcomes, evaluate the likelihood of litigation, and inform legal strategy. AI tools such as ROSS Intelligence and Lex Machina started to provide insights derived from thorough data analysis, giving legal practitioners useful data to aid in their decision-making. Predictive analytics is the process of predicting future events based on past data. AI-driven predictive analytics technologies are used in the legal arena to evaluate historical case data and forecast the probable result of current cases. This program is especially useful for litigation strategy because it helps attorneys foresee potential outcomes and make well- informed judgements based on insights gleaned from data.

6.3.5 AI-Driven Legal Services

AI-driven legal services, like virtual legal assistants, chatbots that generate documents automatically, and automated document creation, have proliferated in recent years. The aforementioned advances are designed to enhance the efficiency and accessibility of legal services, especially for individuals and enterprises who lack the financial means to retain traditional legal counsel.

6.4 Current Applications of AI in Law

AI is being used in the legal industry in a variety of ways to support many facets of legal practice and decision-making. The following are some important areas where AI has had a big impact:

6.4.1 Legal Research and Document Review

Thanks to AI technology, legal research and document evaluation have become much more accurate and efficient. NLP and machine learning-powered tools can swiftly sort through enormous volumes of legal documents, find essential case law, and highlight important details [7]. This lowers the possibility of oversight and mistake while also accelerating the study process.

6.4.2 Case Management and Workflow Automation

Legal processes are automated in a number of ways by AI-driven case management systems, including scheduling, task management, and document organisation. Legal practitioners may increase productivity overall, streamline administrative procedures, and manage their caseloads more skilfully with the use of these solutions [8].

6.4.3 Contract Analysis and Drafting

Legal contract analysis and draughting are being done more and more with AI techniques. Automated systems have the ability to examine contract provisions, spot any problems, and guarantee adherence to legal requirements. This improves contract management accuracy and cuts down on the time and effort needed for contract review.

6.5 Key Milestones and Case Studies

6.5.1 Significant Milestones

- In 1981, the MYCIN expert system was released, and it became a significant model for expert systems in law.
- 1995 saw the introduction of improved AI-driven search features for Westlaw and LexisNexis.
- 2015: IBM's Watson project demonstrated how artificial intelligence (AI) can comprehend and interpret natural language, paving the way for future legal AI applications.

6.5.2 Notable Case Studies

ROSS Intelligence: An AI-driven legal research tool that supports legal enquiries and case studies using NLP [9].

Lex Machina: A platform for legal analytics that offers data-driven insights into legal matters, including intellectual property disputes.

6.6 Impact on the Legal Profession

The legal profession is changing significantly as a result of AI's integration with the area. AI improves access to legal services, accuracy, and efficiency on the one hand. AI-powered solutions, for instance, may manage laborious and repetitive duties, freeing up legal experts to concentrate on more intricate and crucial areas of their work. This change might raise the standard of legal services and increase their accessibility to a wider range of people. However, the growing use of AI in law also brings up significant practical and ethical issues. Critical problems that need to be addressed include data privacy, prejudice in AI systems, and the possibility of legal employment displacement.

Ensuring that AI systems are transparent, accountable, and designed to mitigate biases is essential for maintaining fairness and trust in the legal system.

6.7 Future Directions and Challenges

The future of artificial intelligence (AI) in law is anticipated to be dynamic, with new developments and breakthroughs in technology anticipated. More advanced prediction tools, improved natural language processing, and increased interaction with other technologies like blockchain are possible developments in the field of artificial intelligence in law. But there are further issues with AI adoption that must be resolved [10]. These include resolving ethical issues with the use of AI in decision-making processes, assuring data security and privacy, and creating regulatory frameworks to control the use of AI in legal situations. In order to overcome these obstacles and fully ealize the advantages of artificial intelligence in the legal field, cooperation between scientists, legislators, and attorneys will be essential.

6.7.1 Advancements in AI Technology

Future advancements in AI are likely to further enhance its capabilities in the legal field. Innovations in machine learning and natural language processing may lead to even more accurate predictive tools and efficient legal research systems.

6.7.2 Regulatory and Policy Developments

Regulators and policymakers will need to develop frameworks to address the ethical and legal challenges posed by AI. This includes establishing guidelines for transparency, accountability, and fairness in AI systems.

6.7.3 Integration with Traditional Legal Practices

The future will likely see greater integration of AI with traditional legal practices. This requires legal professionals to adapt to new technologies while maintaining the core principles of justice and ethical practice.

6.8 Challenges and Lessons Learned

6.8.1 Early Limitations

Early AI systems in law faced limitations such as limited data availability, insufficient processing power, and the challenge of accurately replicating human legal reasoning. These challenges led to a cautious approach to AI adoption in the legal field.

6.8.2 Evolution of AI Capabilities

Over time, improvements in data availability, processing power, and AI algorithms have addressed many of these limitations. Modern AI systems are more capable of handling complex legal tasks and providing actionable insights.

6.8.3 Ethical and Regulatory Considerations

The growing use of AI in law has highlighted the need for ethical guidelines and regulatory frameworks to ensure fairness, transparency, and accountability in AI-driven legal decision- making [11].

6.8.4 Predictive Analytics

Predictive analytics uses historical data to forecast future events. In legal contexts, it involves analyzing case outcomes, judicial opinions, and other relevant data to predict the likely result of a case [12].

6.8.5 Case Studies

Platforms like Lex Machina and ROSS Intelligence offer predictive analytics by examining historical case data to identify patterns and trends. For example, Lex Machina provides insights into judicial behaviour, helping legal professionals anticipate case outcomes.Predictive analytics can improve case strategy and efficiency [13]. However, its accuracy depends on the quality of historical data and may be limited by inherent biases in the data.Premonition analyses court records to identify patterns in judicial behaviour, helping legal professionals anticipate how a particular judge might rule on a case.Westlaw Edge utilizes AI to offer insights and recommendations based on vast legal databases, while fastcase provides advanced search capabilities to streamline legal research[14].AI enhances efficiency and accuracy in legal research, but it requires continuous updating and validation to ensure the relevance and correctness of the information provided. AI tools like Westlaw Edge and fastcase leverage natural language processing to assist with legal research and document review. These tools can quickly identify relevant precedents and legal texts.AI systems are used to predict case outcomes based on various factors, including case specifics, judge's past rulings, and opposing counsel's history[16].Outcome prediction can provide valuable insights for case strategy but may be limited by the availability and quality of data and the complexity of human decision-making.AI tools are used to automate contract analysis and drafting, identifying potential issues and inconsistencies.Automation can increase efficiency and accuracy but may miss nuanced legal language or context-specific issues [17]. Impact on the Justice System are as follows:

6.9 Enhancing Efficiency

AI tools streamline administrative tasks, reduce the time required for legal research, and optimize case management. This leads to faster case resolutions and more efficient legal practices.

6.10 Improving Access to Justice

AI-powered platforms provide affordable legal assistance to underserved populations, democratizing access to legal services [18]. Online tools can offer preliminary legal advice and draft documents, reducing the need for expensive legal representation.

6.11 Risk of Bias and Fairness

AI systems, trained on historical data, may perpetuate existing biases. This could lead to biased legal outcomes if not carefully managed [19]. Ensuring fairness requires ongoing evaluation and adjustment of AI systems to mitigate these risks.

6.12 Ethical and Legal Considerations

6.12.1 Accountability and Transparency

AI systems' decision-making processes need to be transparent and accountable. Legal professionals and regulators must establish standards for transparency and hold developers accountable for their systems' performance [20].

6.12.2 Data Privacy and Security

AI tools require access to sensitive personal data, raising concerns about privacy and security. Compliance with data protection regulations and implementation of robust security measures are crucial to safeguarding individual privacy.

6.12.3 The Role of Human Judgment

AI can assist but cannot fully replace human judgment. Legal decisions often involve ethical considerations and contextual factors that AI may not fully grasp. Balancing AI insights with human judgment is essential for fair legal outcomes [21].

6.12.4 Predictive Analytics in Action

A notable case where predictive analytics played a significant role is the use of Lex Machina in intellectual property disputes. By nalyzing past cases, Lex Machina provided insights into potential outcomes, helping firms strategize more effectively.

6.12.5 AI in Document Review

In the case of a major corporate merger, AI tools like Kira Systems were used to review thousands of contracts. The AI's ability to quickly identify key clauses and issues significantly expedited the review process.

6.12.6 Ethical Challenges

In a high-profile discrimination case, an AI system's predictions were scrutinized for bias. The case highlighted the need for continuous monitoring and adjustment to prevent unfair outcomes.

6.13 Conclusion

The role of AI in legal decision-making represents a significant advancement in the legal field, offering opportunities for increased efficiency, improved access to justice, and enhanced legal research. However, the integration of AI also presents challenges related to bias, transparency, and accountability. Ongoing collaboration among legal professionals, technologists, and policymakers is essential to address these challenges and ensure that AI serves to enhance, rather than undermine, the principles of justice.

References

1. C. H. Chien, "Artificial Intelligence and the Law: A Primer," *Harvard Law Review*, vol. 133, no. 2, pp. 389-419, 2020.
2. D. M. Katz and M. J. Bommarito, "Measuring the Complexity of Legal Language," *Stanford Technology Law Review*, vol. 21, no. 1, pp. 1-35, 2018.
3. H. Surden, "Artificial Intelligence and Law: An Overview," *Georgia State University Law Review*, vol. 35, no. 3, pp. 665-680, 2019.
4. M. Westbrook, "Ethical Considerations in AI for Legal Decision-Making," *Yale Law Journal*, vol. 130, no. 5, pp. 1324-1350, 2021.
5. R. Susskind, *Tomorrow's Lawyers: An Introduction to Your Future*. Oxford University Press, 2019.
6. R. Calo, "The Case for a Federal Robotics Commission," *California Law Review*, vol. 103, no. 3, pp. 655-706, 2015.
7. M. Hildebrandt, *Smart Technologies and the End(s) of Law: Novel Entanglements of Law and Technology*. Routledge, 2015.
8. S. Agarwal and M. Reddy, "Artificial Intelligence and the Future of Law: A Review," *Journal of Artificial Intelligence Research*, vol. 62, pp. 445-473, 2018.
9. J. H. Moor, "The Nature, Importance, and Difficulty of Machine Ethics," *IEEE Intelligent Systems*, vol. 21, no. 4, pp. 18-21, 2006.
10. R. Susskind, *The Future of the Professions: How Technology Will Transform the Work of Human Experts*. Oxford University Press, 2017.
11. E. López, "Legal Tech and the Transformation of the Legal Profession: An Analysis," *Journal of Legal Technology Risk Management*, vol. 8, no. 2, pp. 101-115, 2020.
12. R. Binns, "Fairness in Machine Learning: A Survey of Methods and Applications," in *Proceedings of the ACM Conference on Fairness, Accountability, and Transparency*, 2018, pp. 1–11.
13. L. E. Ribstein and B. H. Kobayashi, "Regulating Artificial Intelligence: A New Approach," *Texas Law Review*, vol. 98, no. 4, pp. 903-934, 2020.
14. R. Binns and L. Sweeney, "Discrimination in Machine Learning: A Survey of Emerging Issues," in *Proceedings of the AAAI Conference on Artificial Intelligence*, vol. 32, no. 1, pp. 1-8, 2018.
15. J. Dastin, "Amazon Scraps Secret AI Recruiting Tool That Showed Bias Against Women,"*Reuters*,Oct.10,2018.[Online].Available: https://www.reuters.com/article/us-amazon-com-jobs-automation-insight- idUSKCN1MK08G.
16. T. Nguyen and J. Dastin, "Understanding the Ethical Implications of AI in Legal Practice," *Journal of Ethics and Information Technology*, vol. 21, no. 1, pp. 55-66, 2019.
17. B. Friedman and H. Nissenbaum, "Bias in Computer Systems," *ACM Transactions on Information Systems*, vol. 14, no. 3, pp. 330-347, 1996.
18. J. E. Cohen, "The Regulatory Challenges of Artificial Intelligence in Law," *Harvard Journal of Law & Technology*, vol. 32, no. 2, pp. 420-445, 2019.
19. X. Zeng and L. Zhang, "AI and the Law: An Empirical Study of the Applications of Artificial Intelligence in Legal Systems," *International Journal of Law and Information Technology*, vol. 28, no. 3, pp. 198-215, 2020.
20. J. A. Kroll and S. Barocas, "Accountability in Algorithmic Decision-Making: A Study of the Role of AI in Legal Decision-Making," in *Proceedings of the 2017 ACM Conference on Computer Supported Cooperative Work and Social Computing*, pp. 1025–1036, 2017.
21. A. Santos, "Algorithmic Justice: How AI is Changing Legal Processes and What It Means for the Future of Law," *Journal of Law and Artificial Intelligence*, vol. 5, no. 1, pp. 42-59, 2021.

Chapter 7
The Role of Artificial Intelligence and Machine Learning in Managing Chronic and Mental Health Conditions Through Telemedicine During COVID-19

Poonam and Neera Batra

Abstract Significant changes in healthcare were brought about by the COVID-19 pandemic, particularly in the areas of mental health and chronic disease management. Many experienced financial difficulty, stress, anxiety, and depression. By providing real-time insights, individualized care, and predictive tools to solve these problems, artificial intelligence (AI) and machine learning (ML) enhanced telemedicine. According to a 1,200-person poll, age and workplace culture have a big impact on mental health. This study examines the function of AI and ML during the pandemic and focuses on creating stress prediction and prevention tools utilizing Decision Support Systems (DSS). It emphasizes their uses, difficulties, and potential in healthcare, particularly in the area of mental health, demonstrating how telemedicine and AI and ML may successfully address pandemic issues. This chapter examines how AI & ML, when combined with telemedicine, addressed the specific challenges posed by the COVID-19 pandemic.

Keywords COVID-19 · Artificial intelligence · Machine learning · Mental health · Applications · Challenges

7.1 Introduction

The COVID-19 epidemic has produced significant pressures worldwide by increasing the incidence of depression, posttraumatic stress disorder (PTSD), suicidal tendencies, and anxiety. The absence of predictive biomarkers, the shortage of psychiatrists, and the inherent subjectivity of human judgment all contribute to the global increase in mental health disorders. Therefore, immediate efforts addressing thorough mental health monitoring are required, particularly regarding the early care

Poonam (✉) · N. Batra
Department of Computer Science and Engineering, M. M. Engineering College, MMDU, Mullana, India
e-mail: saini24poonam@gmail.com

© The Author(s), under exclusive license to Springer Nature Switzerland AG 2026

K. Dalip et al. (eds.), *Healthcare 5.0*, https://doi.org/10.1007/978-3-032-09582-4_7

of people exposed to severe anxiety associated with COVID-19. It is of the utmost importance to recognize and identify high-risk individuals when they are still in the early phases of acute stress to reduce the likelihood that these individuals may develop more severe long-term mental health illnesses [1].

Telemedicine, gain popularity and remote solutions during COVID-19 and provide some relief to health care professionals and patients. Collaboration of telemedicine practice with AI & ML, became more effective, more accurate and precise, with great statistical analysis of dieses, decision support system development, patient care from distance became more effective and useful for both practitioners and patients [2]. This chapter delves into how AI & ML contributed to the management of both chronic and mental health conditions during the pandemic, emphasizing the challenges faced and future opportunities. In nutshell, the complete chapter is turning around some research questionnaire:

i. **General Impact:** In what ways has the COVID-19 epidemic highlighted the necessity of ML & AI-powered telemedicine for the treatment of mental and chronic illnesses?
ii. **Technical Assessment:** In comparison to current approaches, how does the suggested ML & AI-based approach enhance performance indicators (accuracy, precision, recall, and F1-score)?
iii. **Patient and Healthcare Provider Perspective:** What issues in mental health telemedicine may AI and machine learning successfully resolve?
iv. **Prospects for the Future:** What developments are required to improve the ethical and dependable application of ML & AI in telemedicine?

7.2 The Role of AI and ML in Telemedicine During COVID-19 a Literature View

ML and AI have the key role in monitoring and patient care without physically visits of health care professionals. Hence, AI and ML may lead health care sector towards telemedicine monitoring system and may offer solution of issues like where in-person dependency and shortage of health care professional. AI-driven telemedicine platforms provided tools for remote monitoring, predictive analytics, and personalized treatment plans tailored to both chronic conditions and mental health needs [3]. ML and AI may provide many decision support systems which may help in identifying individuals at high mental health risk and may predict the parameters that may affect individuals in future also, enabling timely interventions before situation got worsened [4]. Telemedicine's key advantages include:

- Ease of digital transformation in healthcare industry,
- Addressing geographic and social inequalities in healthcare access,
- Enhancing access to medical services through telemedicine practice,
- Improving the efficiency, effectiveness, better clinical solutions of healthcare system,

- Patient engagement in distance monitoring of health, and care coordination across various healthcare services,
- Reducing healthcare costs, etc., these benefits have been well-documented across a range of medical specialties.

The COVID-19 pandemic has drastically affected global mental health, leading to the need for decision-support systems to enhance mental health care. This review covers recent research on these systems, assessing their effectiveness and potential for implementation.

Tutun et al. developed a Decision Support System (DSS) using the Network Pattern Recognition (NEPAR) algorithm to predict mental health status. The system utilizes various machine learning models, achieving 89% accuracy in diagnosing mental disorders [5].

Zhou et al. illustrated the use of machine learning models to predict mental health issues in healthcare workers during public health crises, finding strong results with random forest, gradient-boosting trees, and LASSO models [6].

Akshi et al. used a COVID-19 Mental Health Questionnaire to study psychological stress in Indians and achieved 92.15% accuracy in predicting mental health outcomes using machine learning models [7].

Gupta et al. created a model using XGBoost to predict stress levels based on questionnaire data, reaching 88% accuracy. They identified key stress predictors, such as medicine shortages and concentration difficulties, using SHAP values [8].

Flesia et al. studied metal health stress in 2,053 Italian individuals during the COVID-19, identified 76% sensitivity in analyzing high stress that was closely related to demographic and personality [9].

Abbas et al. used EEG analysis on college students during COVID-19; a model based on LightGBM was developed to discover emotional status, confirming a prominent impact on student mental health [10].

Bhirud et al. proposed PSYCO-19, a chatbot for those facing mental health issues, accomplishing 66% accuracy in identifying English-language texts [11].

Smith et al. arranged a study of 2,000 entries from the Kaggle COVID-19 Mental Health. Dataset was applied with Logistic Regression with the LASSO feature selection. The study focused on some relevant features to identify its impact on mental health status [12].

Nguyen et al. Researchers employed Gradient Boosting along with Recursive Feature Elimination on a sample of 3,500 participants from the Kaggle COVID-19 Mental Health Survey. The study focused on some key relevant features to identify its impact on mental health status [13].

Kim et al. employed 1,500 entries survey dataset from the Kaggle COVID-19 Mental Health, the study applied Decision Tree algorithms along with the Chi-Square feature selection technique. The study focused on some relevant features to identify its impact on mental health status [14].

Singh et al. employed COVID-19 Survey Dataset with 5,000 participants, and applied a Decision Tree model and selected features using the SelectKBest method with $k = 5$. Logistic regression was also used as a supplementary analysis [15].

Sood et al. prepared an extensive study on the COVID-19 Social Media Engagement Dataset with 1,000,000 entries applied Random Forest and SVM models. The study used a Genetic Algorithm for feature selection, with key parameters including population size of 100 and 20 generations [16].

Singh et al. organized a study of 320 students using Random Forest with Recursive Feature Elimination. Focused features included depression, stress, and anxiety, with parameters such as n_estimators = 100 and max_depth = 5 [17].

Li et al. worked on a dataset of 500 entries from the COVID-19 Student dataset. SVM was used for correlation-based Feature Selection. The analysis examined stress and anxiety, setting parameters like C = 1 and kernel = linear [18].

Kim et al. the study employed Logistic Regression with PCA on a sample of 1,000 to analyze depression, stress, and anxiety. Key parameters included C = 0.1 and solver = liblinear [19].

Nguyen et al. proposed a sample of 800 students of the COVID-19 Mental Health dataset, and applied Gradient Boosting with Mutual Information Feature Selection. The primary features analyzed were stress and anxiety, with parameters such as learning_rate = 0.1 and n_estimators = 100 [20].

Chaturvedi et al. proposed a model over 1,182 individuals to assess the impact of COVID-19 on education, mental, and social health. The analysis used the non-parametric Kruskal–Wallis test and the Pearson Chi-Square test to explore relationships between demographic factors and mental health [21].

Peiqing et al. proposed a study on mental health prediction for 2,048 college students using SVM. The model is focusing on factors influencing mental health [22].

Alakuş et al. presented a study on mental health prediction for college students during the COVID-19 pandemic used Gradient Boosting with Lasso Regression on a sample of 5,500 [23].

MSH Mukta et al. researched over 1,290 students with Logistic Regression to predict mental health outcomes, with L1 regularization for feature selection [24].

According to the review presented above, it has been evident that statistical methods and machine learning are essential for the early detection of patients at a high risk of mental health deterioration. Monitoring high-risk COVID-19 patients should be improved using a comprehensive ML-based predictive approach. Therefore, this chapter contributes the following:

- This chapter presents an investigation and analysis of mental health due to the consequences of COVID-19 on the lives of students by designing a support system.
- The Questionnaire was prepared according to personal details before COVID-19, mental health, social health, and work environment.
- This chapter presented statistical tests over collected data. The chapter also depicts the relation between different sub-groups of the Questionnaire.
- Further, different machine-learning approaches are compared to predict mental health status.

7.3 ML and AI-Applications in Managing Chronic and Mental Health During the Pandemic

7.3.1 ML and AI in Tele-Monitoring

Telemedicine includes Telemonitoring as a distance care monitoring system which combines wireless and wearable sensor devices, data transmission, session registration, processing of body parameters, health care management through electronic devices and systems. ML and AI were used to detect/track mental health parameters, such as stress levels and emotional well-being, facilitating timely interventions. This proactive approach reduced the risk of complications by ensuring that patients received timely advice.

7.3.2 ML and AI in Predictive Analytics for Mental Health

ML and AI may perform prediction based evaluation from identified parameters and provide valuable insights about mental health during and after COVID-19 pandemic. ML and AL based statistical analysis enables the professional in finding and prediction of the disease progression, mental health crises, timely Telemonitoring interventions and allowing the practitioners to intervene before condition became critical. AI models took into account factors such as demographics, mental health history, and responses to stress-related surveys, offering insights that helped healthcare professionals personalize interventions, such as therapy sessions or medication adjustments, before conditions worsened.

7.3.3 ML and AI for Tele-Triage (Personalized Care)

ML applications and AI based wearable devices could adapt the management of a patient's medical history, lifestyle, genetics based analysis, and their mental health issues. During COVID-19 the ML and AI predicted and recommended mental health status and personalized interventions, therapy sessions, medical tuning, and current strategies that aligned with patients needs.

7.3.4 ML and AI in Telediagnosis and Tele-Consultation

ML and AI provided Telediagnosis support tools that may detect, predict, mental health issues at earlier stages. For example, AI algorithms analyzing mental health parameters through Telediagnosis may identify early signs of mental health issues, while ML based learning models trained on survey data predicted the mental health, stress, anxiety and depression etc. ML & AI-enabled decision support systems provided real-time interventions and recommendations to professionals, helping them in recognizing high-risk patients in early stage.

7.4 Mental Health Decision Support System Using Machine Learning and Analysis

Research-based responses have been collected to discover the post-COVID-19 consequences of mental health. The data collection for this study belongs to NCR-Delhi students and employees. This dataset includes respondents' demographic details, problems they met during COVID-19, and the effects of COVID-19 on their mental health. The available responses may be used as a training dataset for various machine learning algorithms to develop recent ML-based "stress prediction models." For this research, a web-based survey was conducted on students through Google Forms from December 1, 2022, to January 30, 2023. The online survey questionnaire contained subgroups, as presented in Table 7.1.

This study conducted a cross-sectional survey with a sample size of approximately 1200 students from different demographic details, as in Fig. 7.1.

Figure 7.1 depicts that the gender variable categorizes individuals into three groups: male (1), female (2), and other/prefer not to say (3), with a total of 1218 individuals, a mean of 1.5, and a standard deviation of 0.50. The work area variable categorizes individuals into four sectors with a mean of 2.4 and a standard deviation of 0.93. The age variable represents individuals' ages ranging from 11 to 58 years with a mean age of 18.9 and a standard deviation of 6.28. These graphs collectively illustrate the frequency distributions of gender, work areas, and age to understand the sample population space.

Data from questions Q1–Q12 were analyzed by age, gender, and work area using the non-parametric Kruskal–Wallis test. The validity of the hypothesis is tested by the p-value, which is displayed in Table 7.2. For obtaining the p-value, the null hypothesis is assumed to be true. Statistical significance is indicated by a p-value less than 0.05. The findings indicated that while gender had no effect on the responses, age and employment area did. This test evaluates the association between the variables in the dataset and makes the assumption that the null hypothesis is correct.

Figures 7.2a, b, and c depict the Pearson correlation analysis used to examine the relationship between age categories and various variables. The correlation is evaluated among questionnaire groups.

Figure 7.2a shows the correlation between mental health before and after the COVID-19 outbreak. Q4 and Q5 of the Questionnaire pertain to past experiences with stress, while Q2, Q3, Q6, and Q12 focus on mental health concerns arising after the COVID-19 outbreak. Therefore, six features are selected, and among them, correlation is analyzed. The correlation graph shows that Q4 and Q5 correlate highest

Table 7.1 Questionnaire details used for survey

Questionnaire subgroups	Questionnaires	Responses indicators
Personal details	Name, Email, Gender, Age, State, Work Category 1. How many people currently live (with you) in your home? (Q1)	-
Mental health before COVID-19	2. Have you previously (before COVID-19) been diagnosed with a mental health condition? (Q4)	Yes/No
	3. How stressed did you feel beforeCOVID-19 started? (Q5)	1 = low and 5 = High
Mental health after COVID-19	4. How did you perceive difficulties describing, identifying, and expressing emotions during COVID-19? (Q2)	1 = low and 5 = High
	5. How did you perceive that COVID-19 concerns impact your ability to learn/study? (Q3)	1 = low and 5 = High
	6. How do you perceive the risk of contagion during this period of the COVID-19 pandemic? (Q6)	1 = low and 5 = High
	7. Have you faced any specific challenges during the lockdown period? (Q12)	The score is determined using Carrier Uncertainties, Depression, Anxiety and Panic, Financial Stress and Pressure
Social health	8. How do you perceive the condition of social isolation imposed during this period of the COVID-19 pandemic? (Q7)	1 = low and 5 = High
	9. How do you perceive your relationships with your relatives during theCOVID-19 pandemic? (Q8)	1 = low and 5 = High
	10. How do you perceive the relationships with your colleagues during theCOVID-19 pandemic? (Q9)	1 = low and 5 = High

(continued)

Table 7.1 (continued)

Questionnaire subgroups	Questionnaires	Responses indicators
	11. How do you perceive the relationships with your higher officials (Teachers/professors/boss) during this COVID-19 pandemic? (Q10)	1 = low and 5 = High
Work environment	12. How do you perceive your academic studying experience/work experience during this period of the COVID-19 pandemic? (Q11)	1 = low and 5 = High

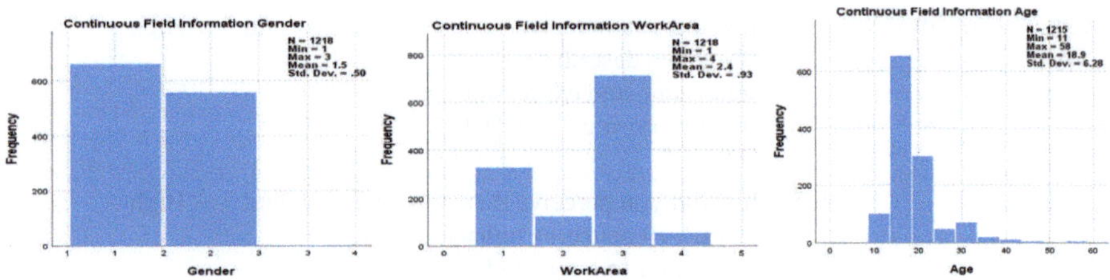

Fig. 7.1 Demographic distribution of data

Table 7.2 Kruskal wallis test summary

Questionnaire	p-value			Questionnaire	p-value		
	Gender	Age	Work area		Gender	Age	Work area
Q1	0.130	0.015	0.276	Q7	0.083	0.002	0.109
Q2	0.858	0.498	0.002	Q8	0.767	0.426	0.244
Q3	0.537	0	0.451	Q9	0.191	0.006	0.018
Q4	0.229	0.443	0.503	Q10	0.180	0.087	0.001
Q5	0.275	0.001	0	Q11	0.591	0.018	0.027
Q6	0.840	0.004	0.113	Q12	0.106	0.002	0.709

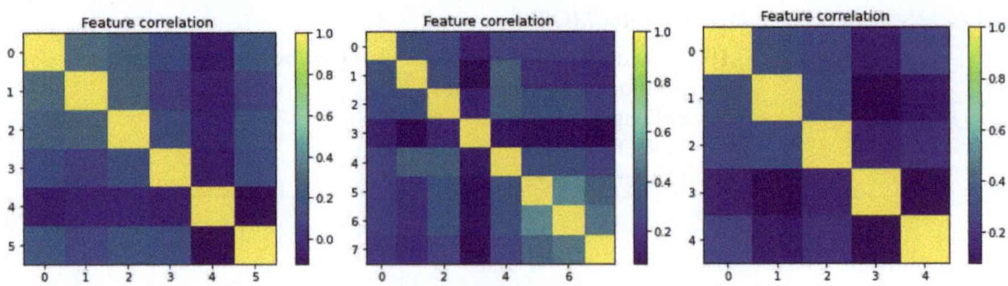

Fig. 7.2 **a, b, c** Correlation analysis among questionnaire subgroups

with Q2, Q3, and Q12. Figure 7.2b shows correlation between social health (Q7, Q8, Q9, and Q10) and mental health after COVID-19 outbreak (Q2, Q3, Q6 and Q12). From this correlation, a total of eight features are selected. The correlation graph shows that Q9 shows the highest correlation with Q2, Q3, Q6, and Q12. Similarly, Fig. 7.2c shows correlation of after COVID-19 outbreak (Q2, Q3, Q6 and Q12) with work environment (Q11). From the correlation graph, it is observed that Q11 shows a correlation with Q2, Q3, and Q12. Therefore, from correlation analysis, it can be concluded that the most contributing questionnaires are Q2, Q3, Q4, Q5, Q9, Q11, and Q12.

The criteria for selecting the features for correlation were based on factors such as mental health before and after the COVID-19 outbreak, social health, and work environment. The selected features are the respective questionnaires from each factor that are hypothesized to have significant relationships with each other to understand the impact of the pandemic on mental health.

Based on the correlation analysis presented, the following conclusions can be drawn:

- Past stress experiences (Q4, Q5) are strongly correlated with mental health concerns during the pandemic (Q2, Q3, Q12).
- Social health (Q9) is significantly correlated with post-pandemic mental health.
- The work environment (Q11) has a substantial impact on mental health during the pandemic.

Figure 7.3 shows the performance evaluation of machine learning for mental health evaluation considering different classifiers. The accuracy of the Naive Bayes classifier is 99%, that of the ANN and SVM is 71%, and that of the logistic regression is 99%. Similarly, the accuracy and recall of random forest are 98%, whereas kNN indicates an accuracy of 75%. This result demonstrates that best result was achieved by Naive Bayes classifier, Logistic regression, and random forest.

This chapter has implemented and trained models in the Keras framework with Tensor Flow, with the training of the proposed model conducted on Google Colab.

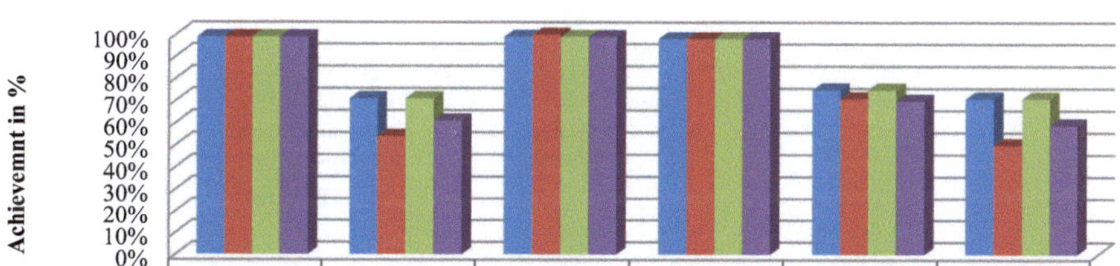

	Naïve Bayes	SVM	Logistic Regression	Random Forest	k-NN	ANN
Accuracy	99%	71%	99%	98%	75%	71%
Precision	99%	54%	100%	98%	71%	50%
Recall	99%	71%	99%	98%	75%	71%
F1_score	99%	61%	99%	98%	70%	59%

Fig. 7.3 Performance evaluation of ML algorithms for mental health evaluation

The collected dataset is divided into 70:30 ratios as training and testing samples for the learning process. Accuracy, Precision, Recall, F1-Score were analysed for evaluation of proposed model.

In Fig. 7.4, the proposed method achieves an average improvement of 16% in accuracy over the existing method [21] (refer Fig. 7.4 for comparative study). In Fig. 7.5, the proposed method achieves an average improvement of 17.5% in precision over the existing method [21] (refer Fig. 7.5 for comparative study).

In Fig. 7.6, the proposed method achieves higher recall with an average improvement of 11.2% over the existing method [21]. In Fig. 7.7, the proposed method achieves an average improvement of 16.5% F1-score over the existing method [21]. Overall, the proposed method performs better for most of the classifiers tested than the existing method.

Fig. 7.4 Comparative accuracy evaluation

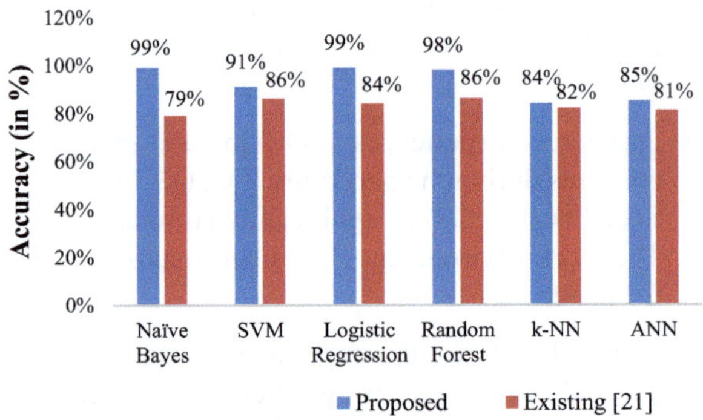

Fig. 7.5 Comparative precision evaluation

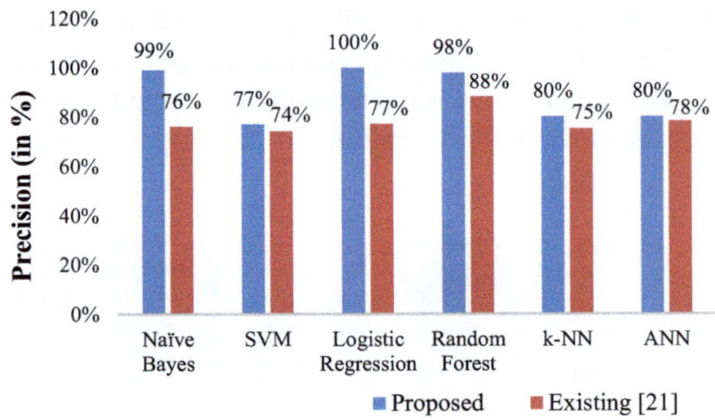

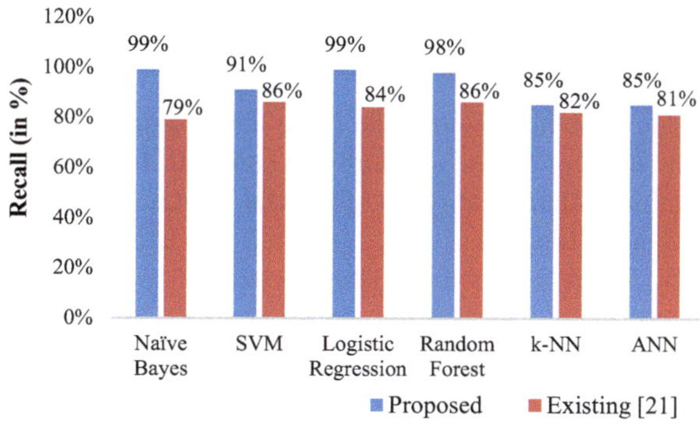

Fig. 7.6 Comparative recall evaluation

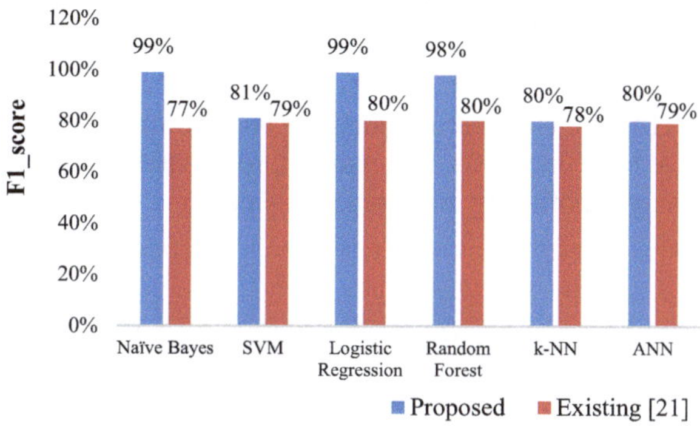

Fig. 7.7 Comparative F1-score evaluation

7.5 Challenges and Limitations

7.5.1 Data Privacy and Security Concerns

Chronic disease and COVID-19 based mental health management requires the continuous collection and communication of individuals information to remotely managed servers which analyses the data that is further used by health care professionals. Ensuring loyalty to data protection is essential to maintain patient confidentiality.

7.5.2 Ethical and Bias Concerns in AI and ML Algorithms

The COVID-19 based dataset may provide biased outcomes and further may lead us to unfair and unethical analysis in critical situations. Addressing these biases is critical to ensuring fair and equitable healthcare delivery, particularly in mental health care, where access to services may already be limited for marginalized populations.

7.5.3 Technical Barriers in AI and ML Implementation

A developing country like India has resource-constrained healthcare infrastructure and lacking behind support to AI and ML technologies. Issues such as data interoperability, data quality, and financial limitations hindered the widespread adoption of AI in healthcare.

7.6 Future Directions

7.6.1 ML and AI-Enabled Virtual Healthcare Assistants (VHAs)

VHAs are enabled in managing and monitoring the symptoms, may intervene with real time diagnosis, and may improvise the treatment management as per AI-driven analysis, offering more reachable care.

7.6.2 Integration of AI, ML and Wearable Technology

ML and AI enabled wearable devices could collect useful data through various health metrices, i.e., stress level, changing habits, etc. and may provide timely diagnostic solutions without requiring the physical presence of the health care professionals.

7.6.3 Personalized Medicine and Genomics in Mental Health

By analyzing genetic, lifestyle, and environmental data, AI can create individualized treatment plans that enhance treatment efficacy while minimizing side effects.

7.7 Discussion

The proposed method outperforms current approaches in accuracy, precision, recall, and F1-score, greatly increasing the efficacy of telemedicine in addressing mental and chronic health issues during COVID-19. Using Naïve Bayes and Logistic Regression, accuracy increased from 79 to 99%, while precision and recall increased by 17.5% and 11.2%, respectively. A balanced performance is highlighted by a 16.5% improvement in the F1-score. These developments improve the identification of important cases, lower the number of incorrect diagnoses, and guarantee accurate forecasts—all of which are essential in times of emergency like COVID-19. In addition to resolving data privacy issues, the study shows how AI and ML can revolutionize telemedicine and opens the door for more research on a variety of datasets and real-time applications.

7.8 Conclusion

The COVID-19 epidemic brought to light how crucial AI and ML are to improving telemedicine for the treatment of mental health issues and chronic illnesses. Though they have drawbacks like algorithmic bias and data privacy, these technologies offer real-time information, predictive analytics, and individualized care. This work demonstrates how machine learning (ML) can be used to identify important aspects of mental stress, providing mental health practitioners with improved diagnostic resources. Notwithstanding sample constraints, further study can improve predictive models and create AI-enabled DSSs, providing a quicker and more affordable substitute for conventional mental health evaluations and revolutionizing medical procedures.

References

1. Dolev, T., Zubedat, S., Brand, Z., Bloch, B., Mader, E., & Avital, A. (2021). Physiological parameters of mental health predict post-traumatic stress in physicians treating COVID-19 patients. *Translational Psychiatry, 11*, 178-185.
2. Topol, E. (2019). *Deep medicine: How artificial intelligence can make healthcare human again.* Basic Books.
3. World Health Organization (WHO). (2021). *Telemedicine: Opportunities and developments in member states.* Global Observatory for eHealth Series 2.
4. Kumar, A., Verma, S., & Goyal, A. (2020). Machine learning for psychological disorder prediction in Indians during the COVID-19 lockdown. *Journal of Medical Systems, 44*(12), 225-232.
5. S.Tutun, M.E. Johnson, A. Ahmed, et al., "An AI-based decision support system for predicting mental health disorders," *InfSyst Front*, vol. 25, pp. 1261–1276, 2023.
6. Zhou, Z., Luo, D., Yang, B. X., & Liu, Z. (2021). Machine learning models for depression symptoms among Chinese healthcare workers during the COVID-19 outbreak. *Frontiers in Psychiatry, 13*, 876995.
7. A. Kumar, "Machine learning for psychological disorder prediction in Indians during COVID-19 nationwide lockdown,"pp. 161–172, 2021.
8. P. Gupta, et al., "Predictive modeling of stress in the healthcare industry during COVID-19: A novel approach using XGBoost, SHAP values, and tree explainer," *IJDSST*, vol.15, no.1, pp.1-20, 2023.
9. L.Flesia, M.Monaro, C.Mazza, V.Fietta, E.Colicino, B.Segatto, P.Roma, "Predicting perceived stress related to the COVID-19 outbreak through stable psychological traits and machine learning models," *J. Clin. Med.*,vol. 9, 3350, 2020.

10. Q. Abbas, A. R. Baig, A.Hussain, "Classification of Post-COVID-19 emotions with residual-based separable convolution networks and EEG signals," Sustainability, vol. 15, 1293, 2023.
11. N.Bhirud, S.Tatale, P.Jain, A.Kulkarni, M.Panchpor, N.K.Jain, "Development of psychiatric COVID-19 chatbot using deep learning,"in: Dey, N. (eds) Data-DrivenApproach for Bio-medical and Healthcare, Data-Intensive Research. Springer, Singapore, 2023.
12. J. Smith, et al., "Predicting mental health in students using machine learning feature selection algorithms,"*International Journal of Machine Learning and Computing*, vol. 10, no. 1, pp. 72-79, 2022.
13. T.Nguyen, et al., "Machine learning feature selection for mental health prediction in students during COVID-19,"*Proceedings of the International Conference on Machine Learning and Data Engineering*, 2022, pp. 113–120.
14. Y. Kim, et al., "A comparative study of feature selection algorithms for mental health prediction in students during COVID-19,"*Proceedings of the International Conference on Data Mining and Machine Learning*, 2022, pp. 65–72.
15. A. Singh, A. Verma, A.Goyal,"Mental health prediction of students during COVID-19 using machine learning," *Proceedings of the 2022 IEEE International Conference on Emerging Trends in Information Technology and Engineering (ICETITE)*, March 2022, pp. 1–6.
16. N. Sood, M. Singh, N.Kumar,"Predicting mental health of students using machine learning techniques during the COVID-19 pandemic," *Proceedings of the IEEE International Conference on Advances in Computing, Communication, and Control (ICAC3)*, March 2022, pp. 1–6.
17. R.Singh, R.Gupta, A.Kaur, "Mental health prediction in students using machine learning feature selection algorithms,"*Journal of Education and Health Promotion*, vol. 10, issue 1, 238, 2021.
18. S.Li, Y.Liu, Y.Yang, "Predicting anxiety in college students during the COVID-19 pandemic: A machine learning approach,"*Frontiers in Psychology*, vol. 12, 659455, 2021.
19. D. H.Kim, S. Y.Kim, S. H.Kim, S. K.Lee, "Machine learning-based prediction of stress in college students during the COVID-19 pandemic,"*International Journal of Environmental Research and Public Health*, vol. 18, issue 3, 1125, 2021.
20. Q. H.Nguyen, T. T.Luu, T. H.Nguyen, D. T.Hoang, "Depression prediction in university students during the COVID-19 pandemic: A machine learning approach,"*International Journal of Environmental Research and Public Health*, vol. 18, issue 6, 2889, 2021.
21. K. Chaturvedi, D. Vishwakarma, N. Singh, "COVID-19 and its impact on education, social life and mental health of students: A Survey,"*Children and Youth Services Review*,vol. 121, 105866, 2020.
22. H. Peiqing, "Multidimensional state data reduction and evaluation of college students' mental health based on SVM," *Journal of Mathematics*,vol. 2022, pp. 1-11, 2022.
23. T. Alakuş, I. Turkoglu, "Comparison of deep learning approaches to predict COVID-19 infection,"*Chaos, Solitons& Fractals*, vol. 140, 110120, 2020.
24. M. S. H. Mukta, S. Islam, S. Shatabda, M. E. Ali, A. Zaman, "Predicting academic performance: Analysis of students' mental health condition from social media interactions," *Behavioral Sciences*, vol. 12, issue 4, 87, 2022.

Poonam a Ph.D. Scholar at the Dept. of Computer Science and Engineering, M. M. Engineering College, MMDU, Mullana, India, her area of research is Machine Learning, Deep Learning, and AI.

Dr. Neera Batra is working as Professor at Maharishi Markandeshwar University, Mullana, specializing in Human-Computer Interaction, Pervasive Computing, IoT, Artificial Intelligence, Machine Learning, Databases, and Distributed Computing.

Chapter 8
Role of Artificial Intelligence in the Healthcare Industry

Varsha Chauhan

Abstract Artificial Intelligence (AI) has had a transformative impact on the healthcare industry in the twenty-first century. This study is based on a comprehensive literature review examining the role of AI in healthcare: clinical diagnosis and diagnostics, virtual patient care, drug development, clinical research, patient engagement and compliance, and clinical care. AI is being used in diagnostic and diagnostic services to evaluate treatment plans to reduce management by healthcare professionals (HCPs). However, this research proposal addresses many intellectual, ethical, and social issues, including privacy, security, policy options and testing, pricing, information and support, access, and performance while integrating medical expertise. The safety and responsibility of the patient and the faith of doctors in the recognition and improvement of health issues depend on the Effective use of AI. Good governance is a must to address ethics, trust and governance issues when promoting the adoption and deployment of AI. The concept of AI has revolutionized healthcare since COVID-19 impacted healthcare systems around the world, and this attack could be another step forward for future medical needs.

Keywords Artificial intelligence (AI) · Healthcare · NLP · Governance

8.1 Introduction

AI has revolutionized the healthcare industry by changing how we identify, treat, and keep track of patients. With the ability to provide more individualized therapies and more accurate diagnoses, this technology is significantly enhancing healthcare research and results. The application of AI in healthcare enables medical personnel to swiftly uncover illness signs and trends that they might otherwise miss by analyzing large volumes of clinical records [1]. AI has a wide range of possible uses in the healthcare industry, from predicting outcomes from electronic health information

V. Chauhan (✉)
MMICT & BM and Research Scholar Department of CSE, MMEC, Maharishi Markandeshwar (Deemed to be University), Mullana, Ambala, India
e-mail: varshacse541@gmail.com

© The Author(s), under exclusive license to Springer Nature Switzerland AG 2026
K. Dalip et al. (eds.), *Healthcare 5.0*, https://doi.org/10.1007/978-3-032-09582-4_8

to analyzing radiological images for early detection. Nowadays, AI can be used in hospital and clinic settings to make healthcare systems faster, smarter, and more effective in treating millions of patients globally. In the field of healthcare, AI is genuinely the wave of the future, revolutionizing the way patients receive high-quality treatment while reducing costs for providers and enhancing patient outcomes [2]. It is quite amazing how artificial intelligence can change the healthcare industry. AI is expected to have a profound impact on how we manage medical data, diagnose diseases, develop treatments, and put an end to them. AI in healthcare enables medical personnel to make better judgments based on more precise information, which saves time, lowers expenses, and enhances overall medical records administration.

AI in healthcare promises to be a game-changer, paving the way for a time when patients receive high-quality care and treatment more quickly and correctly than in the past, from discovering novel cancer treatments to enhancing patient experiences [3]. AI in healthcare makes it possible to process and evaluate enormous volumes of medical data significantly more quickly than is possible for humans. This capacity was crucial for making treatment recommendations, forecasting results, and identifying illnesses. For example, AI algorithms are faster and more accurate than human radiologists in analyzing medical pictures, such as MRIs and X-rays. They can frequently identify diseases like cancer early on [4].

Artificial intelligence (AI) in healthcare can forecast possible health concerns by examining trends in a patient's past medical history and present health data. Healthcare professionals can give proactive, preventative care thanks to these predictive capabilities, which eventually improve patient outcomes and lowers healthcare expenditures [5].Healthcare applications can use Robotic Process Automation (RPA).Some of these responsibilities include medical records, income management, claims processing, and medical records management [6]. Healthcare organizations (HCOs) are also using chatbots for patient communication, mental health, and telemedicine [7]. The scope of this field in healthcare can improve patient care outcomes by enabling decisions to be made faster and increasing the accuracy of diagnoses. This is accomplished through more seamless workflows. Thanks to the power of artificial Intelligence, Doctors can work quickly and efficiently on complex data, allowing for timely intervention and informed treatment plans. Capacity to create skill-based, customized treatment plans. It will reduce side effects and provide better treatment. Thanks to the results, new drugs and treatments can be approved more quickly. This is accomplished through more seamless workflows [8].

8.1.1 Literature Survey

Diagnostics

Medical imaging diagnosis has benefited from AI's pattern-recognition capabilities, particularly in machine learning and deep learning. For example, by accurately evaluating radiographic pictures, convolutional neural networks (CNNs) can identify conditions including cancer, heart disease, and neurological illnesses [9].

Personalized Medicine

To determine which treatments will work best for a certain patient, AI systems examine genetic information, medical records, and treatment results. Personalized medicine is being advanced by this predictive method, which enables customized treatment strategies for improved results and fewer adverse effects [10].

Drug Discovery

Through the analysis of large datasets, therapeutic efficacy prediction, and clinical trial candidate identification, artificial intelligence (AI) speeds up the drug discovery process. In order to save time and money, machine learning techniques are being employed to model how medications interact with biological systems [11].

8.1.2 Medical Imaging

AI's computer vision algorithms assist radiologists with more accurate image analysis and follow-up suggestions by interpreting MRI, CT, and X-ray pictures and detecting anomalies [12].

Disease Prevention and Predictive Analytics

AI predicts illness risks, readmissions to hospitals, and outbreaks by analyzing patient data and electronic health records (EHRs). Proactive healthcare management is made possible by these insights, which aid in resource planning and early interventions [13].

Robotics in Surgery

AI-guided robotic devices improve patient recovery times, decrease invasiveness, and increase surgical precision. By assisting surgeons with intricate procedures, these devices lower human error and enhance overall results [14].

Administrative Automation

Administrative efficiency is made possible by the automation of healthcare billing, scheduling, and documentation by natural language processing (NLP). Additionally, AI facilitates patient follow-up and triage procedures, freeing up medical personnel to concentrate on vital patient care duties tasks [15].

8.2 Analysis and Evaluation of Ai Machines

Analysis and Evaluation AI Machines: can analyze medical images such as X-rays, MRIs, and CT scans to identify abnormalities, help diagnose diseases, and monitor vital signs. Artificial intelligence uses deep learning models, specifically convolutional neural networks (CNN), to process and interpret complex data. This model is trained from a large database of medical records. Using cognitive science and art is a multi-step process that includes data collection, model training, imaging, and treatment [16]. Below are specific expert practices in this field.

8.2.1 Data Collection and Preparation

Data Collection

Imaging data: From hospitals, clinics, research centers, etc., collection of medical images (X-rays, MRI, CT scans). Surgeons often collect these images to highlight areas of interest, areas of suspicion, or specific conditions, providing baseline facts for training AI models. Pixel usage and compatibility mode.

Training Model Algorithm Selection

A Deep Learning Model: Convolutional Neural Networks (CNN) are frequently used image analysis models because they can learn the hierarchical features of data. Early models of large-scale public information systems (such as ImageNet) are often fine-tuned to use clinical data to supplement existing knowledge and improve performance. The model used to train a CNN learns to map input images to corresponding diagnoses or classifications.

Image Processing and Analysis Feature Extraction

Hierarchical Approach: CNN extracts features at different levels, from simple edges and textures in the first layer to patterns and patterns in deep layers.

Force mechanism: Certain sophisticated models employ a method to highlight the primary image, thereby increasing the accuracy of diagnosis).

8.2.2 Clinical Integration

Decision Support: AI as an Assistant: AI systems provide decision support to radiologists by highlighting potential concerns and recommending diagnoses. Intelligence can act as the radiologist's second opinion, ensuring safety and reducing the likelihood of misdiagnosis. Seam access AI is analyzed with traditional image data.

8.2.3 Continuous Improvement and Validation

Post-Deployment Monitoring

Monitoring: Continuously monitor the performance of your AI models in a real environment to ensure accuracy and reliability. At the same time, the ideas of electricians and doctors are used to improve and develop the model. Use historical safety and effective medical practices.

Radiology: Smart models detect lung, bone, and other abnormalities through X-rays and CT scans to improve early diagnosis effects. Eye diseases. By doing this, AI can enhance results for patients by increasing diagnosis accuracy and streamlining the diagnostic process [17].

8.3 Predictive Analysis

Role: AI can predict outcomes of patient treatment, disease progression, and health risks to ensure health and identity. Model: Algorithms analyze patient history data, including demographics, medical history, and treatment outcomes, to identify conditions and predict future conditions. Remove important information from here. The system uses a variety of machine-learning techniques to predict results for patients, infection rates, and health risks. Detailed information on how artificial intelligence is used in this field is given below:

Demographic information, test results) and unnecessary information (e.g., medical information). **Wearable Devices:** Continuous health monitoring data such as heart rate, activity level, and sleep patterns.

Imaging: Information from imaging studies that can be combined with other medical information. Please make sure the measurement of the difference is different. (NLP):

Use NLP technology to extract information from unstructured data.

8.3.1 Training Model Algorithm Selection

Tracking Learning: Generally used for prediction tasks where the probability variable (such as disease occurrence) is known. Algorithms include linear regression, decision trees, random forests, support vector machines (SVM), and neural networks. Techniques include k-means clustering and auto encoders.:

The data set is divided into two: The training set and the validation set. The training method is used to train the model, and the reference method is used to tune parameters and prevent over fitting.:

Use methods such as grid search or random search to improve the algorithm to improve the performance of the model [18].

8.3.2 Model Validation and Evaluation

Performance Metrics

Accuracy: Proportion of correct predictions for all patients. Important. Compromise on speed. Subset to ensure robustness.

8.3.3 Deployment and Integration

Clinical Integration

Deploy Into EHR Systems: Integrate predictive models into electronic medical records for instant risk assessment testing and decision. The worksheet provides predictions and insights for practitioners. Monitor performance standards and update standards as new information becomes available. Examples of Predictive Analytics in Healthcare.

Disease Risk Prediction: Artificial intelligence models predict the likelihood of diseases such as diabetes, heart disease, and cancer based on patient data. Pressure, so make a response plan to prevent this. Thus creating a self-healing system. Using advanced technology, AI can provide valuable information that improves patient care, improves outcomes, and increases treatment benefits [19].

8.3.4 Personalized Medicine

Role: Artificial intelligence prepares treatment plans for patients based on their cosmetic, lifestyle and health needs information. Intelligent analysis of genetic information to identify mutations and other biomarkers that affect disease and treatment. Artificial Intelligence in Personalized Medicine involves the use of advanced techniques and large amounts of data to tailor treatments and interventions to individual patients. Detailed information about the use of artificial intelligence in this field is given below:

- Biomarker data. Includes medical history, diagnosis, treatment, test results, and medical records.
- Details regarding the patients' dietary and exercise habits as well as the effects they have on the environment. Keep a close eye on data like blood sugar levels, heart rate, exercise level, and other vital signs. Constructing the patient's biological profile.

8.3.5 Data Preprocessing and Analysis

Preprocessing Steps

Standardization: Standardize data to ensure consistency across different data types and locations.

Data Quality

Feature Engineering: Create new features from raw data to capture better data for predictive models. It can predict health outcomes and treatments. And principal component analysis (PCA) to identify patient groups and reduce data complexity.

Model Development Algorithm Selection

Supervised Learning: A model that predicts specific outcomes based on training data. Common algorithms include decision trees, random forests, gradient-boosting machines, and neural networks.

Reinforcement Learning: A model that learns the best strategy through repeated trials can be used for adaptive strategies.

8.3.6 Predictive Modeling and Personalization

Risk Prediction

Disease Risk Score: Calculates an individual's disease risk score based on genetic, clinical and lifestyle data. Models to predict diseases such as cancer, diabetes and heart disease.

Optimum Dose: Determine the optimal dose for each patient to maximize effectiveness and minimize side effects.

Convenient treatment plan: It is not necessary to regularly adjust treatment plans based on immediate data and patient response.

8.3.7 Clinical Medicine

Decision Support System

EHR Integration: Embed intelligent models into electronic medical records to provide physicians with truly real-time decision-making. Doctors give advice on self-treatment and risk assessment. Apps: Create mobile apps that provide patients with personalized health tips, alerts and monitoring tools. Continuous Learning and Improvement Feedback:

Live Data: Patient data is continuously collected and reviewed to refine and improve clinical practice. Accuracy and effectiveness of artificial intelligence. Research: AI models identify genetic changes and biomarkers to recommend personalized cancer treatments and predict patient response to immunotherapy. Customized protection strategies. It helps manage chronic conditions such as diabetes. By integrating and analyzing disparate data, intelligence can predict disease risk, personalize treatment, and continually improve outcomes for patient. This approach increases the accuracy and effectiveness of treatment, leading to better health and better care [20].

8.4 Robotic Surgery

Role: Artificial intelligence-powered robots assist surgeons by increasing accuracy, reducing deviations, and reducing complications during surgery.

Computer Vision: Artificial Intelligence systems use computer vision to guide surgical instruments with high efficiency. Precision and control; results from professional algorithms, real-time data analysis and machine learning. You can find detailed information about how artificial intelligence is used in robotic surgery here.

Detailed 3D model of the patient's body. The intelligence-driven software simulates multiple surgical procedures to optimize surgical planning.

Analytics: AI predicts potential complications and outcomes based on patient data and past surgeries [21].

8.4.1 Assistance During Surgery Introducing Time

Surgical images: Artificial Intelligence provides real-time image updates by combining preoperative images with surgical videos in a timely manner and delivering them to the surgeon. Intelligence-driven AR improves accuracy by relaying relevant information into the surgeon's view.

Self-Service*:* AI can perform simple, repetitive tasks, such as suturing or using tissue, under the supervision of a surgeon. Improved accuracy and safety.

8.4.2 Improved Accuracy

Micro Motion: Intelligent control robots can perform precise micro movements beyond human ability, important for small jobs. Machine learning algorithms instantly adapt to surgical conditions, always accurate.: Artificial intelligence provides tactile feedback to surgeons, suggesting that the amount of force required from robotic tools should be used to prevent tissue damage. Scientific Research and Further Studies Scientific Research:

Data Collection: AI collects data during surgery, including video footage, measurement measurements, and patient vital signs. Provide feedback on surgical performance and identify areas for improvement.:

Artificial intelligence-supported simulators provide realistic training to surgeons using data obtained from real surgeries. Artificial Intelligence Examples in Robotic Surgery Systems **DaVinci Surgical System Artificial Intelligence Features:** This widely used robotic system uses artificial intelligence to improve precision, scale, and reduce shakiness.: It is used in many surgeries, including urology, gynecology, and heart surgery.

Real-Time Analysis: The system provides real-time analysis and adjusts the surgical plan as needed. Do a biopsy. Looking Forward to the Future Enhanced Autonomy:

AI-Assisted Autonomy: Future advancements may allow AI to operate more autonomously, with a surgeon controlling all procedures. Artificial intelligence systems will increase their capabilities and reliability by continuously learning from each process.

A Big Data Analytics: Using big data to focus on improving surgical procedures and predicting outcomes. Artificial intelligence is most effective in the field of robotic surgery by using the best images, real-time data analysis, machine learning and continuous learning of data. This makes the surgical procedure more precise, efficient, and safe, making it beneficial for patients and doctors [22].

8.5 Virtual Health Assistant

Role: AI-powered chatbots and virtual assistants improve access to medical services by providing medical information, alerts, and support to patients.

Natural Language Processing (NLP): NLP enables these assistants to understand and answer patient questions in natural language. The use of artificial intelligence in virtual healthcare assistants (VHA) increases patient engagement, communicates medical information, supports doctors in daily tasks, and provides personal health education. Here is a detailed discussion of.

How AI is being integrated into virtual healthcare:

Speaking in written language. Advanced NLP algorithms (such as Google's Speech-to-Text or Amazon's Transcribe) can accurately record user speech. Thinking, rooting and lemmatization. A model such as GPT-4 can produce consistent and appropriate responses.

8.5.1 Personalized Health Coaching Data Integration

Electronic Health Record (EHR): VHA integrates with the EHR to access the patient's medical history, medication lists, and laboratory results to assist in making recommendations for the individual.

Tools: Data from health and social care providers provide information on key times, activity levels, and sleep patterns. A certificate of symptoms stating the nature of the condition and recommending further study.

8.5.2 Automate Daily Tasks and Scheduling

Calendar Integration: AI assistant can access and manage users' calendars, schedule medical appointments, and send reminders. Patients send messages to check on their progress.: Artificial intelligence can alert patients to drug interactions based on their current medication list.

8.5.3 Depression Support Interviewers

Interactive Conversations: AI-powered VHA uses technology to simulate interactive conversations, providing support in focusing thinking and reducing feelings of loneliness. Some VHAs offer CBT practices and mood tracking to support mental health. A doctor or emergency room.

8.5.4 Patient Education Medical Information Search:

Evidence-Based Guides: VHA provides information from reputable medical institutions to ensure patients receive accurate medical information and are updated. Situations and perceptions interact with various dimensions of health Validity [23].

8.5.5 Data Privacy and Security Secure Data Processing

Encryption: All patient information processed by VHA is encrypted to prevent unauthorized access. HIPAA USA ensures the privacy and security of patient information.

8.5.6 Examples of Intelligence-Powered Virtual Health Assistants

Ada Health

Symptom Checker: Guides the user through step-by-step guidance before using artificial intelligence to evaluate and diagnose symptoms.

Integrating data to provide health advice: Using machine learning to improve the accuracy of health assessments.

Artificial Intelligence: Allows users to participate in social conversations and motivates by encouraging conversations, supporting mental health and educating patients. Using advanced NLP, data integration, machine learning, and secure data processing, AI-driven VHA can improve healthcare, performance [24].

8.6 Drug Discovery and Development

Role: Artificial intelligence accelerates the process of discovery and development of new drugs, reducing the associated time and costs. It also finds how compounds interact with targets in the body to identify successful candidates for further testing. AI is playing a revolutionary role in drug discovery and development by accelerating all stages, from initial target identification to clinical trials. Detailed information on how artificial intelligence is used in this field is given below:

Identify targets where new drugs can target genes or proteins. Techniques such as deep learning can reveal complex patterns in gene expression data. Biomarkers of patient response to treatment to aid target validation.

8.6.1 Drug Design and Optimization Molecular Modeling

Model-Based Drug Design: Artificial intelligence uses techniques such as molecular docking to predict how small molecules will interact with target proteins. Tools like Deep Mind's Alpha Fold can predict protein structures with high accuracy and guide the design of potential drugs.

Optimization

ADMET Prediction: Artificial intelligence predicts the absorption, distribution, metabolism, excretion and toxicity (ADMET) characteristics of potential drug users, optimizing them more efficiently and safely.

(QSAR): Artificial intelligence models help improve manufacturing processes by predicting the biological activity of compounds based on their chemical structure.

High Through Put Screening Virtual Screening

Chemical Libraries: AIs screen large chemical libraries to identify potential therapeutic compounds. Machine learning models rank compounds based on their predictive performance against targets.

Pheno Typic Analysis

Image Analysis: Artificial intelligence analyzes cell images through high-throughput analysis to identify phenotypic changes caused by compounds and reveal therapeutic effects.

Preclinical Testing Toxicity Prediction

In Silico Models: Artificial intelligence models predict the toxicity of compounds early in the drug development process, reducing reliance on animal testing. This model uses historical data to identify potential negative outcomes.

(PK/PD): Simulation model: Artificial intelligence simulates pharmacokinetics (how drugs are absorbed, distributed, metabolized and excreted) and pharmacodynamics (effects of drugs in the body) or Predicted uptake and side effects.

Clinical Trials Patient Recruitment

Predictive Modeling: Artificial Intelligence analyzes electronic medical records, genomic data, and social media to identify suitable candidates for clinical trials. This creates more competition between patients and trials, resulting in better performance. It also monitors data in real-time during testing to identify critical conditions and adjust processes as needed.

Self-Healing: AI identifies patient groups most likely to benefit from treatment and promotes self-healing.

Post-Marketing Monitoring Discovery Findings

Pharma Covigilance: Artificial intelligence monitors real-world data, including electronic medical records and social media, to identify adverse drug reactions and ensure continued safety. Learning algorithms enable risk management by detecting possible negative effects that may arise from big data. Predicting the binding of small molecules to target proteins and accelerating identification.

Knowledge Graph: Use artificial intelligence to analyze large amounts of biomedical data to identify new and repurposed drug targets. Discovery and development. Using advanced algorithms and big data, AI has transformed the traditional pharmaceutical manufacturing process, making it faster, more efficient, and more precise, ultimately resulting in better outcomes and patient care [25].

8.7 Administrative Task

Responsibilities: AI automates routine tasks such as scheduling, billing, and managing patient information, allowing medical professionals to focus on patient care.

Automation tools: AI uses robotic process automation (RPA) to solve repetitive tasks. And we continue to evolve. By leveraging the power of advanced algorithms and big data, AI can increase diagnostic accuracy, personalize treatment, enhance efficiency and patient results. The potential for AI to transform healthcare will increase as technology advances, leading to new and improved outcomes, reduced costs, and increased efficiency. Below is detailed information on how AI can be integrated into various administrative functions:

Used by Natural Language Processing (NLP) to interact with patients to manage appointments and cancellations. They can handle multiple demands simultaneously and optimize schedules to reduce absenteeism and manage hospital operations efficiently.

8.7.1 Medical Billing and Coding Automated Coding

Natural Language Processing (NLP): Artificial Intelligence extracts relevant information from medical records and assigns medical codes (ICD-10, CPT) for billing. This eases the burden on human coders and minimizes errors.

8.7.2 Support Services

Security Detection: Intelligent tools analyze payment patterns and detect anomalies that may indicate fraud, ensuring management respect and reducing financial losses. Ability to work to resolve emerging issues and improve claim approval.

8.7.3 Registration and Patient Information Data Entry Automation

Optical Character Recognition (OCR): Artificial Intelligence-powered OCR technology makes it easier to enter patient information into the electronic health record (EHR) by converting written or printed text into text). To guarantee the information's completeness and correctness. As a result, there is less administrative labor and a decrease in human entry errors.

8.7.4 *Inventory Management and Services Inventory Tracking*

Real-Time Monitoring: Intelligent tools monitor inventory in real time and track the usage of repair supplies, illnesses, medications, and equipment. It plans or processes to forecast future product needs to ensure critical supplies are available. Reduces the risk of out-of-stock and overstocking by reducing orders.

8.7.5 *Human Resources Management Recruiting and Hiring*

AI Analytics: AI resume analysis tools and applications to identify the most qualified candidates based on prioritization criteria. They can also use chatbots or voice assistants to conduct initial interviews.

Optimization Algorithms: AI plans changes and manages work distribution to ensure the best results, including staff availability, skills and maintenance. Personal training and education development recommendations to increase employee satisfaction and retention.

8.7.6 *Revenue Management Revenue Management*

Scientific Analytics: AI predicts patient payment behavior and identifies risk factors income to ensure good management.: Analysis of billing, payments, and historical accounts increases revenue and reduces administration [26].

8.7.7 *Patient Communication and Engagement Virtual Assistants*

ChatBots: AI chatbots provide information to patients, answer frequently asked questions, and assist administrators, such as updating personal information or understanding billing.

Communication: AI analyzes patient data to deliver personalized messages and reminders for appointments, medication adherence, and follow-up care.

8.7.8 Compliance Management and Reporting Compliance Monitoring

Real-Time Alerts: Intelligent systems monitor activities and generate real-time reports for any activity that may violate the policy to be monitored.

Automated Reporting: AI automatically generates tracking reports, reducing the time and effort required to manually collect data. Automate repetitive tasks in healthcare management, such as pre-authorization, eligibility checks, and application processing.

Schedule Appointments: Zocdoc's AI-powered platform allows patients to find doctors online and book appointments, shortening appointment times and reducing workload. Information and assistance with medical codes and accurate billing reduce costs and increase accuracy. By streamlining workflows, optimizing resource allocation, and providing insights, AI can help doctors focus more on patient care and reduce administrative burdens. This can increase efficiency, improve patient experience, and improve health [27].

8.8 Result and Discussion

AI is working to bring better health time data to improve everything from wait times to patient transport. Conversational AI has made significant progress in the development of chat bots that can assist patients, monitor their progress, and provide personal health related information. Some of these services can visit patients at home and, after assessing the patient's information and diagnosing the symptoms, recommend treatment or make an appointment with a specialist. AIin health care has the potential to improve patient outcomes, enable the development of health care professionals, and spur advancements in the workforce. The above technology is designed to facilitate diagnosis and diagnosis, combat infection, provide virtual patient care, increase people's participation in pain and treatment according to the law, reduce health management, promote innovation in drugs and vaccines, enable patient tracking and monitoring. Compliance with exercise and gait analysis for technology assisted rehabilitation. As AI advances in healthcare, it will have to overcome many technical, ethical and regulatory challenges. It has raised concerns about data security and privacy due to its use of private and sensitive data that is restricted by equal rights. The current quality of health data and AI's inability to replicate certain human characteristics, such as emotions, may limit AI's ability to solve problems. AI has its advantages when it comes to efficiency, but it cannot change the ties between the people who make up the group. Because robots cannot relate to humans, they cannot perform human collaboration, team management, and other tasks. Analyzing whether AI can be developed and used in a way that is based on human interests and considers technical, ethical, and social issues will be one of

Table 8.1 Healthcare professionals [32]

Diagnosis and treatment	AI-powered tools analyze medical data, such as imaging, pathology reports, and genetic information, to assist in diagnosing diseases like cancer and predicting patient outcomes. These systems provide faster and often more accurate results than traditional methods
Drug discovery	AI accelerates drug development by analyzing large datasets to identify potential drug candidates and predict their efficacy, reducing costs and time-to-market
Personalized medicine	AI enables tailored treatment plans by analyzing patient-specific data, including genetic makeup, lifestyle, and medical history
Administrative efficiency	AI streamlines administrative tasks, such as scheduling, billing, and medical documentation, allowing healthcare providers to focus more on patient care

the most important problems in the management of future intelligence technology. This study expands the existing literature on the use of AI in clinical and diagnostic, virtual patient care, drug discovery, and clinical research, engaged patient care and monitoring, treatment, and other management areas. Additionally, this is the most recent update of the literature on the intellectual, ethical, social, and managerial issues healthcare professionals face when using AI in treatment as shown in Table 8.1.

With an emphasis on accounting, business and management, decision sciences, and health profession studies, this study attempts to present a bibliometric analysis of publications on AI in healthcare.

Second, this paper finds AI the healthcare sector is revolutionized by artificial intelligence (AI), which improves accessibility, accuracy, and efficiency. In order to help diagnose illnesses like cancer and forecast patient outcomes, AI-powered technologies evaluate medical data, including imaging, pathology reports, and genetic information. Compared to conventional techniques, these technologies yield findings more quickly and frequently with greater accuracy.

Third, this paper guarantees the effective and well-coordinated provision of healthcare services and frees up healthcare providers to concentrate more on direct patient care. Lastly, co-occurrence offers some intriguing facts regarding keywords [33].

References

1. Maleki Varnosfaderani, Shiva, and Mohamad Forouzanfar. "The role of AI in hospitals and clinics: transforming healthcare in the 21st century." *Bioengineering* 11, no. 4 (2024): 337.
2. Artificial intelligence in medical imaging: Moving from radiographic pathological data to clinically useful endpoints (Oren O., 2020; Gersh B.J., Bhatt D.L.). 2:e486–e488. https://doi.org/10.1016/S2589-7500(20)30160-6. Lancet Digit. Health.
3. Wang Y., Dong Q., Shen H., Wang Y., F., Jiang Y., Zhi H., Dong Y., Li H., Ma S., Wang Y., Wang Y. 2017;2:230–243. Healthcare with artificial intelligence: History, present, and future. Neurology of Stroke Vascular. https://doi.org/10.1136/svn-2017-000101.

4. 2013; 16:1–8; Dilsizian S.E., Siegel E.L. Using Big Data and Advanced Computing to Provide Personalized Medical Diagnosis and Treatment: Artificial Intelligence in Medicine and Cardiac Imaging. Curr. Cardiol. Rep.
5. Rossi S., Trunfio M. 2022; 1:103â “129. Metaverse Research Advances: Research Streams and Future Agenda. Online environments.
6. Commins J. (2010). Nurses. Distractions Reduce Time Spent at Bedside by 25%. Leaders in Health
7. 2018. Utermohlen K. Four Applications of Robotic Process Automation (RPA) in the Healthcare Sector, Medium.
8. Am J Phys Med Rehabil. 2019;98:e128â“e129; Anderson D. PM&R Applications of Artificial Intelligence.
9. Litjens, Geert, Thijs Kooi, Babak Ehteshami Bejnordi, Arnaud Arindra Adiyoso Setio, Francesco Ciompi, Mohsen Ghafoorian, Jeroen Awm Van Der Laak, Bram Van Ginneken, and Clara I. Sánchez. "A survey on deep learning in medical image analysis." *Medical image analysis* 42 (2017): 60–88.
10. Topol, Eric J. "High-performance medicine: the convergence of human and artificial intelligence." *Nature medicine* 25, no. 1 (2019): 44-56.
11. Vamathevan, Jessica, Dominic Clark, Paul Czodrowski, Ian Dunham, Edgardo Ferran, George Lee, Bin Li et al. "Applications of machine learning in drug discovery and development." *Nature reviews Drug discovery* 18, no. 6 (2019): 463–477.
12. Shen, Dinggang, Guorong Wu, and Heung-Il Suk. "Deep learning in medical image analysis." *Annual review of biomedical engineering* 19, no. 1 (2017): 221-248
13. Rajkomar, Alvin, Jeffrey Dean, and Isaac Kohane. "Machine learning in medicine." *New England Journal of Medicine* 380, no. 14 (2019): 1347-1358.
14. Liu, Hanxiao, Karen Simonyan, and Yiming Yang. "Darts: Differentiable architecture search." *arXiv preprint* arXiv:1806.09055 (2018).
15. Davenport, Thomas, and Ravi Kalakota. "The potential for artificial intelligence in healthcare." *Future healthcare journal* 6, no. 2 (2019): 94-98.
16. Lee SI, Celik S, Logsdon BA, et al. Nat Commun 2018;9:42. A machine learning approach to integrate big data for precision medicine in acute myeloid leukemia.
17. B.Z. Hameed et al. 13 (2021), Artificial intelligence (AI) in engineering and medicine: machine learning and data science developments paving the way for future Article 17562872211044880; Ther. Adv. Urol.
18. ChatGPT in radiology: the benefits and constraints of artificial intelligence for medical imaging diagnosis, S. Srivastav et al., 15 (7) (2023). Cureus
19. Fut. Sci. OA, 6 (2), p. FSO439; T.O. Tobore (2019), On the necessity of developing a cancer early detection, diagnostic, prognostic, and therapy response system
20. Precision Medicine, Artificial Intelligence, and the Future of Personalized Medicine, K. B. Johnson et al., 2021.86Ã—93 in Clin. Trans. Science, 14 (1)
21. 2020 Standards and Classifications in Robotics, Heidegger T. Springer. In: Eng MH, editors, Khatib O, Siciliano B. Terminology in Robotics. Germany: pages 1â“10.
22. 2019;270(2):2 Panesar S, Cagle Y, Chander D, Morey J, Fernandez-Miranda J, Kliot M. Robotic surgery using AI in the future. Anne Sug.
23. In 2023, Arora A, Palmer J, Alderman JE et al. published 29(11):2929â“2938. The importance of standards for health data sets in AI applications. night.
24. Privacy-preserving artificial intelligence in healthcare: methods and applications (N. Khalid et al.Biomed. Comput. Sci. 2023; Article 106848
25. Establishment World Health Organization, W.H. (2017). Novel insights on worldwide health expenditures for universal healthcare. Tech. Rep.
26. In 2019, Beierle, C., and Kern-Isberner, G. Regelbasierte Systeme. In: Methoden wissensbasierter Systeme, edited by C. Beierle and G. Kern-Isberner. Springer.
27. N. Aoki (2020). an experimental investigation into public perceptions of AI chatbots in government. 37(4), 101490; Government Information Quarterly.

28. precision therapy for acute myeloid leukemia by combining big data and machine learning, Logsdon BA, Lee SI, Celik S, et al. Nat Commun 2018;9:42.
29. Artificial intelligence (AI) in engineering and medicine: advances in machine learning and data science opening doors for future research, B.Z. Hameed et al. 13 (2021); Ther. Adv. Urol.
30. https://www.researchgate.net/publication/350949853_Position_paper_of_the_EACVI_and_EANM_on_artificial_intelligence_applications_in_multimodality_cardiovascular_imaging_using_SPECTCT_PETCT_and_cardiac_CT/figures
31. Alderman JE, Palmer J, Arora A, et al. published 29(11):2929â“2938 in 2023. The significance of standards in AI-driven applications for health datasets. Nat Med.
32. McNulty, Cliodna AM, Barry D. Cookson, and Michael AO Lewis. “Education of healthcare professionals and the public.“ *Journal of antimicrobial chemotherapy* 67, no. suppl_1 (2012): i11-i18.
33. Alowais, Shuroug A., Sahar S. Alghamdi, Nada Alsuhebany, Tariq Alqahtani, Abdulrahman I. Alshaya, Sumaya N. Almohareb, Atheer Aldairem et al. “Revolutionizing healthcare: the role of artificial intelligence in clinical practice.“ *BMC medical education* 23, no. 1 (2023): 689.

Chapter 9
Healthcare 5.0: Transforming Telemedicine Workflow Through Innovation

Nikita Sharma

Abstract The emergence of Healthcare 5.0 represents a paradigm shift in the delivery of healthcare services, focusing on the integration of advanced technologies to enhance patient-centric care. This chapter explores the transformative potential of Healthcare 5.0 in telemedicine workflows, highlighting innovative approaches that streamline processes and improve health outcomes. By leveraging artificial intelligence, big data analytics, Internet of Things (IoT), and blockchain technology, telemedicine can be revolutionized to provide personalized, efficient, and secure healthcare experiences. We analyze current challenges in telemedicine, including data privacy, interoperability, and patient engagement, and propose innovative solutions that harness the capabilities of Healthcare 5.0. This study emphasizes the need for a collaborative approach involving healthcare providers, technology developers, and policymakers to create a robust telemedicine ecosystem. The findings suggest that embracing the principles of Healthcare 5.0 can significantly enhance telemedicine's effectiveness, accessibility, and sustainability, ultimately leading to improved patient outcomes and a more resilient healthcare system.

Keywords Healthcare 5.0 · Telemedicine · Internet of Medical Things (IoMT) · Blockchain in telemedicine · Robotics in surgery · Telehealth · Cyber security in healthcare

9.1 Introduction

The integration of innovation in healthcare can significantly enhance patient outcomes, streamline workflows, and improve overall system efficiency [1]. Emerging digital technologies continues to evolve posing unprecedented opportunities in health systems globally to improve healthcare services delivery. There has been significant progress in healthcare [2].

N. Sharma (✉)
Department of Computer Science and Engineering, Poornima University, Jaipur, India
e-mail: nikita.sharma@poornima.edu.in

© The Author(s), under exclusive license to Springer Nature Switzerland AG 2026

K. Dalip et al. (eds.), *Healthcare 5.0*, https://doi.org/10.1007/978-3-032-09582-4_9

A new paradigm known as "Healthcare 5.0" encourages unprecedented cooperation between physicians, patients, and machines by using technology to both support and enhance human competency [2]. The arrival of Healthcare 5.0, which combines patient-centric care techniques with cutting-edge technology like blockchain, IoT, robotics, and artificial intelligence, is causing a paradigm shift in the healthcare Industry [3]. Healthcare 5.0's personalized medicine enables medical treatments to be tailored to a patient's unique genetic composition and medical background, leading to more efficient treatments and improved patient outcomes. Healthcare 5.0 also provides more advanced diagnostic tools with wearable sensors and AI algorithms, allowing for earlier disease detection and improved disease management. Thus, Healthcare 5.0 uses cutting-edge technology to revolutionize patient care, healthcare services, technicians, and the general patient experience, resulting in major advantages for the healthcare industry [4, 5]. Since the introduction of artificial intelligence, it has become standard practice to use AI applications to aid in disease diagnosis, enhance disease prognosis, and shorten patient treatment times. Fifth-generation communication is required as the fundamental network architecture in Healthcare 5.0 in order to connect healthcare devices. IoT will generate data that AI can use, advancing digital wellness by focusing not just on the quality of life and well-being of patients but also on the well-being and well-being of people globally. The main problems with healthcare 4.0 are information leakage and perfect information transfer. Two cutting-edge healthcare 5.0 technologies that have the power to completely change any kind of work are automation and artificial intelligence. Healthcare 5.0's definition of intelligence includes concepts like virtual patient monitoring and detection, remote surgery, accurate and methodical disease diagnosis, and intelligent therapy, which includes online education for anxious patients [6, 7]. The term artificial intelligence (AI) refers to a wide range of sophisticated technological developments. It speaks to the capacity of machine learning techniques to forecast outcomes without the need for human intervention. The phrase "smart health system" was created as a result of the remarkable growth and development of technology in the field of medicine [8].

A key component of Healthcare 5.0 is the requirement for a patient-centric approach to care, which involves tailoring interventions and treatments to each patient's unique needs. Reducing the use of unnecessary tests and treatments can help lower the number of resources used in healthcare as a whole [5]. Technology can be helpful, especially when it comes to connecting with far-flung areas and providing higher-quality care at a more affordable price. Telemedicine is the practice of diagnosing and treating patients from a distance using communication networks. Payers become increasingly aware of how much less expensive telemedicine is than traditional treatment, and physicians become more aware of its advantages, as more patients have access to it. A more advantageous technology that may improve long-term health is telemedicine, which can increase access to preventive care. The field of telemedicine has enormous potential to impact public health [9]. According to the American Telemedicine Association (ATA), telemedicine is an electronic medium of connecting patients and doctors who are located in different places and exchanging medical information [10].

9.2 Healthcare's Evolution

The 1.4 billion people who live in India are served by a broad range of medical services through the intricate and varied network of public and private sectors that make up the country's healthcare system. The system still faces several difficulties even after going through major modifications over time. Among these difficulties are a lack of healthcare experts, insufficient money for public healthcare, discrepancies between urban and rural areas, poor infrastructure, and a fragmented healthcare system. The number of non-communicable diseases in India is rising, which presents a serious obstacle to the country's healthcare system [11].

9.2.1 Healthcare 1.0 to Healthcare 5.0

Healthcare 1.0 (1970 to 1990): Laying the Groundwork

Healthcare 1.0 is the initial phase of contemporary healthcare systems, laying the groundwork for the formalization of medical care. The best way to characterize healthcare 1.0 is as a physician-centric approach. In this case, the doctor was the main source of medical information and was crucial to the decision-making process regarding healthcare [12]. The patient, who frequently knew very little about medicine, had complete faith in the doctor's knowledge. Physicians largely made all of the choices regarding healthcare; individuals had little said in how they were treated. Despite its simplicity, this approach had some drawbacks. Errors and inefficiencies resulted from a reliance on manual operations and non-standard record-keeping [2]. Moreover, the lack of preventive care frequently resulted in expensive therapies and late-stage diagnosis.

Healthcare 2.0 (1991 to 2005): Embracing Technology

Healthcare2.0 introduced new medical technology and digital devices to improve patient care, signaling a dramatic turn toward healthcare industrialization [13]. During this time, electronic health (e-health) was developed, electronic medical devices were adopted, and digital technology for patient imaging and tracking was just getting started. These developments greatly enhanced diagnosis, treatment, and overall healthcare efficiency. Patients gained more control over their health as they started to access internet health information [2]. The emergence of health insurance during this time period also contributed to the tripartite system that includes payers, patients, and providers. The use of imaging test equipment (such as MRI, ultrasound, and CT scan), monitoring devices (such as arterial lines and pulse oximeters), and surgical and life support equipment (such as da Vinci robots and chest tubes) is on the rise in hospitals.

Healthcare 3.0 (2006 to 2015): The Rise of Electronic Health Records (EHRs)

Healthcare 3.0 is defined by automation and sophisticated information systems. Electronic Health Records (EHRs) have revolutionized the healthcare industry and made it possible to provide more effective treatment and remote healthcare capabilities [14]. In response to the demand for a more individualized and comprehensive approach, Healthcare 3.0 was born. This stage made use of developments in data analytics, genetics, and connected devices to customize patient care. Additionally, telehealth and remote care are now feasible through the use of existing computer networks, and certain in-person interactions are starting to give way to electronic visits (such as communication between a patient and their doctor via a patient portal). The development of telemedicine made it easier for people, especially those living in rural places, to obtain healthcare treatments. Health systems became more interoperable, which made it possible to share patient data across platforms with ease.

Healthcare 4.0 (2016 to 2019): Smart Systems and AI-Driven Healthcare

Healthcare 4.0 refers to the improved connectivity between digital and physical elements, as well as the solutions for these connections offered by cutting-edge ICTs like cloud computing, Big Data, and the Internet of Things (IoT) [15]. In such a scenario, the process of delivering health care turns into a cyber-physical system that is outfitted with wearables, RFID (radio frequency identification), the Internet of Things, and various medical devices, intelligent sensors, medical robots, etc. These are combined with cloud computing, big data analysis, artificial intelligence, and decision support techniques to create a smart and connected health care delivery system. Along with the health care organizations and facilities (such as hospitals, clinics, and long-term care facilities), this system also connects all of the gadgets and equipment, as well as the homes and communities of the patients. It may be possible to communicate patient-related data through appropriate protocols, including as prescription history, diagnostic notes, test findings, treatment plans, pharmacy refills, billing, and insurance claims [14]. Additionally, we might imagine proactive therapy, illness prevention and prediction, tailored medicine, and improved patient-centered care with AI techniques. Healthcare 4.0 successfully raises the standard, adaptability, efficiency, affordability, and reliability of healthcare services while also improving the whole healthcare experience.

Healthcare 5.0 (2020 to Present): The Future of Human-Centric Healthcare

Healthcare 5.0 offers wellness monitoring and control, ambient assisted living, emotive telemedicine, smart self-management, smart treatment reminders, compliance and adherence, patient remote monitoring, tracking, and virtual clinics, as well as personalized and connected health care [2]. Sensors are utilized all around the world to enhance AI, mobile devices, nanotechnology, virtual and augmented reality, and transportation. From their inception, sensor technologies have greatly enhanced healthcare. Health data has been captured and converted into detectable electrical impulses using sensors. Health information, including blood pressure, temperature, weight, oxygen saturation rate, stress level, heart rate, blood sugar, and blood pressure, is typically recorded via sensory smart devices and sent as electrical pulses for additional processing. In particular, during the COVID-19 pandemic, sensors are opening up new avenues for virtual care delivery and are continuing to revolutionize health systems around the world. Smartphones, smart wearables, and IoMT have all effectively incorporated sensors with the ability to remotely collect and process health data. Building on the strengths of Healthcare 5.0, the future may include a system that offers a comprehensive approach to patient wellbeing by fusing technology and human interaction. This will prioritize preventive care while managing illnesses and promoting general well-being (Fig. 9.1).

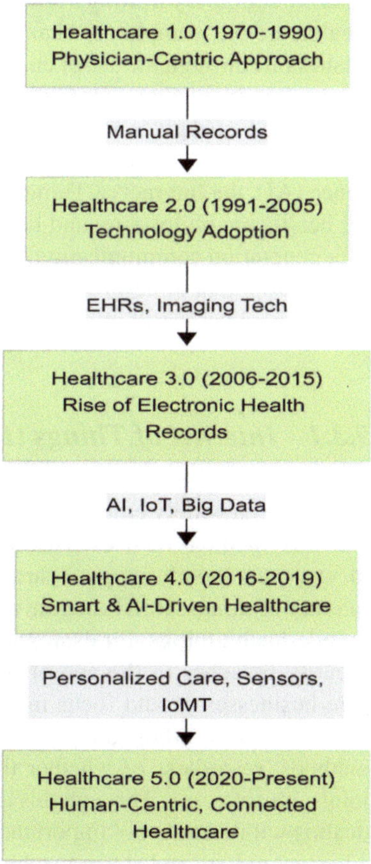

Fig. 9.1 Evolution of healthcare from Healthcare 1.0 to Healthcare 5.0

With each phase being progressively shorter, Healthcare 1.0 (from 1970 to 1990), Healthcare 2.0 (from 1991 to 2005), Healthcare 3.0 (from 2006 to 2015), Healthcare 4.0 (from 2016 to 2019), and Healthcare 5.0 (from 2020 to the present) illustrates the rapid advancement of healthcare innovation and paradigm shifts that have occurred over the last 50 years. The shift from a paradigm where information was viewed selectively to a digital paradigm where information is more easily accessible has had a significant role in the quick development of the digital ages. The evolution of computers, from mainframes to minicomputers and finally to personal computers, marking this progress, culminates in the portability of iPads, which facilitates information access.

9.3 Key Elements of Healthcare 5.0

As information technology has advanced, the concept of healthcare 5.0 has quickly gained traction. By making healthcare more efficient, convenient, and individualized, Healthcare 5.0 completely changes the traditional medical system [16]. It accomplishes this through the use of cutting-edge information technologies including cloud computing, blockchain, artificial intelligence, big data analytics, and the Internet of Things (IoT). [6] gives a high-level summary of all the core concepts that make high-speed data transmission networking approaches possible, including artificial intelligence (AI), the Internet of Things (IoT), and 5G communication. Set a new standard for healthcare philosophy and healthcare facilities. It highlights how crucial 5G, or fifth generation communication, infrastructure, embedded AI, and IoT devices are to the development of new technologies in healthcare facilities and how improvements in people's lives result in healthier lives.

9.3.1 Internet of Things (IoT)

IoT has been widely employed in the healthcare industry to share health data online and link medical devices. Healthcare sensor-based IoT is evolving in a number of ways, each with unique characteristics. The internet of health things, wearable internet of things, IoMT, cognitive internet of medical things, IoNT, and internet of mobile-health things, for example, are all made possible by IoT. In order to prevent security breaches in the age of the Internet of Healthcare Things (IoHT), healthcare businesses should focus more on implementing safe cryptosystems [17]. The IoMT provides orthopedic patients with every treatment conceivable in a variety of methods, regardless of whether the issue concerns the bones, tendons, ligaments, joints, muscles, etc. Through this difficult COVID-19 period, orthopedic patients are dealing with a number of important and serious problems.3, 4 With the help of cutting-edge technology and clever machine learning-based techniques, the suggested IoMT concept provides answers and treatments for these orthopedic patient-related problems. This is especially beneficial in light of the current COVID-19 pandemic's backdrop [18]. One such technology that actively ensures Quality of Service (QoS) criteria is the Cognitive Internet of Things (CIoT), which allows all physical entities worldwide to communicate and exchange information. Machine-to-machine communication in a growing number of wireless device networks is supported by the Cognitive Radio (CR) enabled IoT, or CIoT [19]. The idea of traditional healthcare is being replaced by smart healthcare, which is made possible by IoT and smart IoT devices [6].

9.3.2 Artificial Intelligence (AI)

When AI is used in healthcare facilities, the massive data sets produced by these facilities can be processed using various machine learning algorithms to do forecasting and analysis for medical purposes. Analyzing a patient's sample and symptoms to get a diagnosis from a large dataset of each patient is extremely challenging for lab technicians and medical professionals. Which, in contrast to lab staff, can be handled efficiently by a variety of machine and deep learning algorithms with a lower error rate and more accuracy. Large volumes of data gathered from various smart IoT devices may be processed quickly by AI learning algorithms, which can then forecast the outcome and generate electronic health reports (ERHs) that are sent to the appropriate medical departments for additional review and recommendations. AI systems are capable of acting like supercomputers and thinking like people. It can more accurately and quickly identify sickness. AI can help patients with potential illnesses that could lead to major health issues in the future by using their EHR history to recommend potential care, treatments, and drugs [20]. Furthermore, early diagnosis of diseases like cancer and cardiovascular disorders can be facilitated by AI-powered systems, opening a crucial window for prompt intervention. AI is capable of predicting possible issues, suggesting preventive strategies to reduce future risks, and recommending individualized care regimens by utilizing past data from electronic health records. Additionally, AI chatbots and virtual health assistants improve patient involvement by offering medication reminders, post-treatment care instructions, and real-time consultations, guaranteeing all-encompassing healthcare support. By lowering human effort, minimizing diagnostic delays, and assisting in resource optimization, AI's incorporation into healthcare also greatly raises the general effectiveness and caliber of medical services.

9.3.3 Block Chain (BC)

Healthcare 5.0 faces a number of problems and difficulties, such as handling massive data quantities, a lack of standards, data security risks, regulatory difficulties, etc. In recent years, advocates of smart healthcare have been using the word "blockchain", and a number of research studies have examined the potential applications of blockchain technology in this area. The decentralized and encrypted database where patient data is kept guards against manipulation and unwanted access. Furthermore, it facilitates the safe and transparent exchange of information between patients and healthcare providers. Moreover, it ensures the confidentiality of sensitive data and promotes interdisciplinary cooperation. Blockchain addresses important issues like data breaches and illegal access by ensuring safe, transparent, and impenetrable data management in the healthcare industry. Blockchain is perfect for preserving the integrity of electronic health records (EHRs) because it generates a decentralized ledger that keeps patient records in an unchangeable format [21]. By controlling who can access their data, patients can protect their privacy and ensure that laws like GDPR and HIPAA are followed [22]. Blockchain reduces inefficiencies in care coordination by enabling smooth data sharing between labs, hospitals, and insurers. By monitoring medicine sources, eliminating counterfeiting, and guaranteeing quality, it improves the pharmaceutical supply chain [23]. Blockchain technology in clinical research guarantees the transparency and integrity of trial data, boosting confidence in the results. Processes like consent management and insurance claims are automated using smart contracts, which lowers errors and administrative burdens. Real-time, tamper-proof data interchange is ensured via integration with IoMT devices, improving precision medicine and remote monitoring. The use of blockchain in Healthcare 5.0 is an example of a safe, patient-focused, and effective system for the healthcare industry of the future [24].

9.3.4 Telemedicine Adoption

In addition to exposing the dispersion of healthcare delivery systems, the COVID-19 pandemic reinforced the need for robust primary healthcare networks in order to improve public health responses during health emergencies [25]. The general public's initial point of contact with the healthcare system is primary care, although money and attention have not increased recently for this service. Even with the COVID-19 pandemic, telemedicine presents an opportunity to overcome long-standing obstacles to primary care in India, including a lack of trained physicians, access problems, and the high expense of in-person care. The implementation of universal health coverage might be accelerated while primary care was reinforced by telemedicine. As defined by the World Health Organization (WHO), telemedicine is" the delivery of medical services, where distance is an essential aspect, by all health care professionals using technology for communication and information for the exchange of valid information for study and evaluation, diagnosis, treatment, and prevention of disease and injuries, as well as for the continuing education of health care providers, all in the interests of advancing the health of individuals and their communities" [25]. According to the American Telemedicine Association (ATA), telemedicine is an electronic means of connecting patients and doctors who are located in different places and exchanging medical information [10].

Palliative care, early detection, improved treatment, prevention, and rehabilitation are all aided by telemedicine in the context of cancer care. The health systems at secondary and tertiary care facilities are overburdened as a result of a lack of primary care delivery networks and referral units. Adequate planning and operational procedures are necessary for the successful use of telemedicine. Therefore, the advancement and application of telemedicine will enhance patient care as well as the primary healthcare system in India in the future. Ultimately, the most important result of telemedicine will probably be how cost-effective it is.

9.4 Case Studies in Healthcare 5.0 Telemedicine

Healthcare 5.0 offers wellness monitoring and control, ambient assisted living, emotive telemedicine, smart self-management, smart treatment reminders, compliance and adherence, patient remote monitoring, tracking, and virtual clinics, as well as personalized and connected health care (Fig. 9.2).

Case Study 1: Telemedicine for Chronic Disease Management Globally, the primary causes of death are chronic conditions like diabetes, high blood pressure, and heart disease. The conventional healthcare 1.0 method of treating chronic disorders sometimes necessitates repeated in-person visits to medical professionals, which can be difficult and time-consuming for patients. The utilization of telemedicine consultation and telemonitoring techniques improved the management of rheumatoid arthritis, diabetes, and hypertension [10]. Wearable ECG monitors and smartphone apps have made it possible to remotely monitor, diagnose, and treat heart diseases, significantly improving telecardiology techniques. These solutions reduce the need for frequent hospital visits for patients with heart disease by enabling continuous cardiac health tracking, smooth data transmission to healthcare providers, and timely feedback [26]. A telemedicine initiative launched by a healthcare provider, HealthTrack, aimed to improve chronic disease management through remote patient monitoring (RPM) and teleconsultations, aligning with Healthcare 5.0's goal of integrating advanced technologies for personalized care. Healthcare systems are investigating cutting-edge strategies like RPM and telemedicine to improve the efficiency, affordability, and accessibility of managing chronic diseases. Telemedicine has been shown in earlier research to improve patient outcomes, especially when paired with ongoing monitoring.

Diabetes, hypertension, and rheumatoid arthritis were all better managed by telemedicine, particularly when telemonitoring and consultation were employed. The ideal duration of intervention for individuals with diabetes using telemedicine as a disease management strategy is 12 months. In individuals with rheumatoid

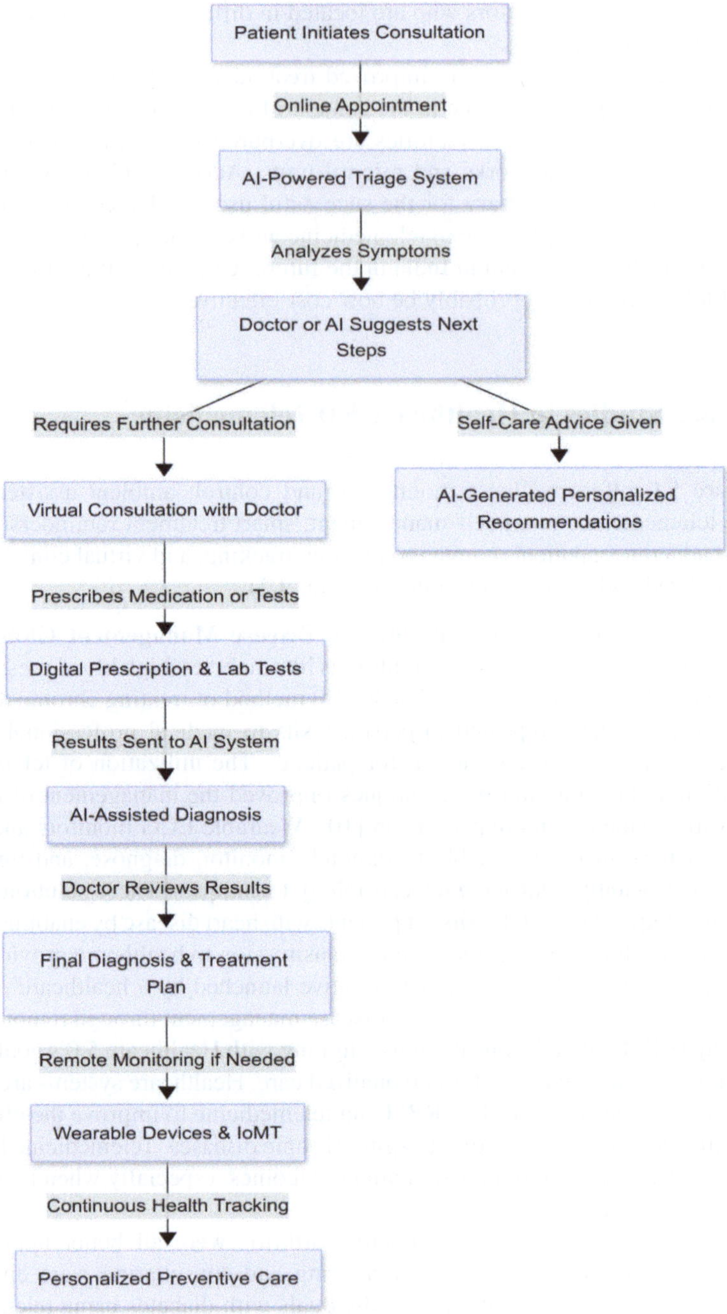

Fig. 9.2 Telemedicine workflow in Healthcare 5.0

arthritis, telemedicine enhanced medication adherence and decreased negative feelings, whereas in hypertensive patients, it lowered systolic blood pressure [10]. Integrating telemedicine with data analytics platforms and electronic health records (EHRs) guarantees smooth communication and individualized treatment, enabling medical practitioners to make well-informed decisions at a distance. AI also makes predictive analysis possible in telemedicine, providing customized interventions and improving the overall management of chronic illnesses, resulting in better patient outcomes and lower healthcare expenses.

Case Study 2: Telemedicine Advancing Remote Surgical Care A key component of Healthcare 5.0, remote surgical support uses telemedicine and state-of-the-art technologies to revolutionize the way surgical care is delivered, removing regional boundaries and providing access to worldwide knowledge. The advent of telemedicine, which makes remote surgery possible, has fundamentally altered the way that doctors practice. Telemedicine's growth enabled telesurgery, the most important use of telemedicine equipment, by making it simpler to remotely monitor patients [27]. Leading the way are robotic surgical devices, which enable skilled surgeons to carry out complex operations from a distance. When combined with haptic feedback, these solutions guarantee control and accuracy while simulating the tactile sensation of conventional surgery. By building immersive, virtual operating theaters where surgeons may view patient anatomy in three dimensions, interact with real-time data, and work together with other medical experts globally, the Metaverse improves on this strategy.

A groundbreaking development in Healthcare 5.0, metaverse-assisted telesurgery offers revolutionary remote treatment options. Blockchain and XAI work together to guarantee safe, open, and effective surgical procedures, showing promise for transforming healthcare around the world [28]. Healthcare 5.0 technologies enable telesurgery, which is a major advancement in the provision of surgical care. A novel approach was created to guarantee accuracy, openness, and confidence in remote surgeries by combining the Metaverse, Blockchain, and Explainable AI (XAI). This case study highlights the revolutionary potential of these technologies by examining how they worked together to provide successful remote surgical operations. A surgeon in the United States used a virtual operating room (created in the Metaverse) to perform laparoscopic surgery on a patient in India. Real-time haptic feedback devices ensured precision, while a 3D view enhanced situational awareness. Telemedicine in healthcare 5.0 has the potential to transform remote surgical care. By integrating 5G, AR, and robotics, it tackles the dual issues of quality control and geographic constraints. Additionally, research highlights the need of blockchain integration for safe patient data handling, promoting regulatory compliance and trust [29].

The objectives of Healthcare 5.0, which include providing individualized, safe, and effective care, are in line with remote surgical help, which combines telemedicine, robots, blockchain, and XAI. In addition to improving accessibility and surgical accuracy, this paradigm change also tackles issues of data security and trust, establishing a new benchmark for the provision of healthcare globally in the digital era.

Case Study 3: AI-Enhanced Telemedicine for Mental Health Healthcare 5.0's AI-driven telemedicine is transforming individualized mental health treatment by combining cutting-edge algorithms, real-time data analytics, and virtual platforms to provide easily accessible and customized mental health services. Obstacles such as stigma, lack of mental health experts, and geographic restrictions frequently affect traditional mental health care. Through the provision of distant, effective, and flexible therapeutic interventions, AI-powered telemedicine tackles these issues [30]. Globally, mental health problems including stress, anxiety, and depression have become more prevalent, especially in neglected and rural areas where mental health services are hard to come by. In accordance with Healthcare 5.0 ideals, a healthcare firm called MindWell revolutionized care by delivering customized, stigma-free mental health services through wearable technology and AI-driven telemedicine.

AI-driven chatbots offer round-the-clock assistance, bridging the time between planned therapy sessions by giving mindfulness exercises or cognitive behavioral therapy (CBT) procedures [31]. During virtual therapy sessions, artificial intelligence (AI) systems use natural language processing (NLP) to evaluate the speech patterns, tone, and sentiment of patients. This allows them to diagnose disorders such as depression, anxiety, or post-traumatic stress disorder and provides important insights into emotional states [32]. A virtual treatment platform might, for example, identify tiny indications of emotional distress in a patient's voice and suggest prompt follow-ups. In order to provide individualized and dynamic care methods, machine learning models dynamically modify treatment plans in response to patient progress. Blockchain promotes system confidence by guaranteeing the safe and private handling of patient data.

By providing a customized, ongoing, and comprehensive experience, this method not only improves mental health care but also increases accessibility for marginalized groups. Through the integration of wearable technology, telemedicine, and artificial intelligence, Healthcare 5.0 is revolutionizing mental health treatment, empowering patients, and enhancing results worldwide.

9.5 Emerging Trends Influencing the Future of Telemedicine in Healthcare 5.0

The merging of cutting-edge technologies like artificial intelligence (AI), the Internet of Things (IoT), blockchain, virtual reality (VR), augmented reality (AR), and robotics will propel the revolutionary future of telemedicine in Healthcare 5.0. Telemedicine has a bright and revolutionary future thanks to technological breakthroughs, shifting patient preferences, and changing healthcare demands. Telemedicine processes could be redefined and healthcare delivery improved in a number of ways by emerging trends and innovations.

9.5.1 Artificial Intelligence and Machine Learning Integration

The delivery of healthcare services and patient care will be completely transformed by the incorporation of AI in telemedicine. Virtual health assistants, who offer 24/7 patient monitoring and prompt medical advice for common illnesses, can be powered by AI. By evaluating patient data and adjusting treatment regimens appropriately, it can also aid in the management of chronic illnesses. By analyzing health data, AI can help with early disease identification, enabling preventive care. AI is a potent tool for telemedicine patient outcomes improvement and healthcare personalization because of its capacity for learning and adaptation.

Additionally, machine learning models are increasing clinical efficiency, decreasing administrative tasks, and enhancing automation of workflows. Algorithms driven by machine learning examine patient symptoms, medical records, and real-time data to rank cases according to urgency. This increases the effectiveness of care by guaranteeing that urgent patients receive attention as soon as possible. AI-powered solutions are revolutionizing telemedicine workflows by fusing sophisticated algorithms with user-centered designs to provide more accurate, patient-focused, and efficient procedures.

9.5.2 Expansion of Remote Patient Monitoring

One revolutionary development in telemedicine is remote patient monitoring, or RPM. It enables prompt interventions and lessens the need for hospital stays by enabling healthcare providers to track patient health data in real-time [33].

Through the use of digital technology, RPM gathers health information from patients in one place and securely sends it to medical professionals in other places for evaluation. RPM will be essential in controlling chronic diseases, enhancing health outcomes, and cutting healthcare costs as the industry shifts to a more preventive paradigm. Continuous monitoring of patients' health is made possible by RPM systems, which enable early identification of abnormalities or worsening conditions. For instance, high blood pressure or irregular cardiac rhythms may cause prompt intervention signals.

Healthcare professionals can improve treatment outcomes by developing highly customized care plans that are suited to each patient's needs thanks to the insights obtained by RPM systems. In order to ensure prompt care, advanced RPM systems can link with telemedicine platforms, enabling medical professionals to monitor patients from a distance and hold virtual consultations.

9.5.3 Rise of Wearable Technology

Another significant trend impacting telemedicine is wearable technology. Vital health metrics including heart rate, blood pressure, oxygen levels, and glucose levels are tracked by contemporary wearables like smartwatches, fitness bands, and biosensors. Real-time data from these gadgets gives patients and medical professionals useful information.

By offering a constant stream of health information, these gadgets can enable patients to take a more active part in managing their own health. Additionally, they can help in remote patient monitoring, which allows medical professionals to monitor patients' health and take appropriate action. Preventive care and tailored treatment will be improved by the combination of wearable technologies and telemedicine. These gadgets deliver real-time data that gives patients and medical professionals useful insights that improve health outcomes and management Sikander et al. [34].

9.5.4 Emerging Virtual and Augmented Reality (AR/VR)

The immersive and interactive tools that virtual reality (VR) and augmented reality (AR) provide are transforming telemedicine by improving several facets of healthcare delivery. AR superimposes important diagnostic information and visual aids to enhance comprehension during consultations, while VR creates immersive settings that allow healthcare professionals to engage with patients in real-time, facilitating virtual consultations. The remote training of healthcare professionals is another important application for these technologies. By providing real-time instruction during complex surgeries, AR ensures precision and increases trust, while VR simulations enable practitioners to rehearse medical procedures in risk-free, realistic environments.

Additionally, these technologies enable worldwide remote cooperation, enabling real-time consultation and expertise sharing among healthcare practitioners. Additionally, they remove geographical restrictions, enabling underserved and distant locations to obtain cutting-edge treatment.

9.5.5 Blockchain for Secure Health Transaction

There is great potential for blockchain technology in the field of telemedicine. It can improve interoperability, patient privacy, and data security; all of which are crucial telemedicine concerns. Blockchain can produce a decentralized, safe patient data record that can only be accessed with the patient's permission. By making it easier for authorized clinicians to securely share health information, this can enhance patient outcomes and care coordination.

Additionally, it can offer an open audit trail, boosting confidence in telemedicine services. Overcoming the difficulties of data fragmentation and interoperability in telemedicine depends on the technology's capacity to improve data integrity, security, and efficiency. Blockchain thereby promotes openness and trust, both of which are essential for telemedicine's broad acceptance and success.

Through seamless connectivity and individualized treatment, healthcare 5.0 combines smart technologies with human-centered healthcare to improve patient results and experience. Telemedicine will develop into a comprehensive, patient centered, proactive, and predictive system of treatment in this new era, moving beyond straightforward consultations. In general, telemedicine in Healthcare 5.0 will prioritize providing individualized, effective, and easily accessible care, hence enhancing the proactive, intelligent, and integrated nature of healthcare. By bridging the gaps in healthcare delivery, this digital transformation promises to improve outcomes, lower costs, and make high-quality treatment available anywhere, at any time.

9.6 Challenges in Telemedicine

Telemedicine, as an integral component of Healthcare 5.0, presents various challenges that must be addressed to unlock its full potential. The table below outlines the key challenges faced in telemedicine, their possible solutions, and the resulting impacts that can drive the transformation of healthcare systems globally (Table 9.1).

By addressing these challenges with strategic solutions, telemedicine can play a pivotal role in the transformation of healthcare under Healthcare 5.0. The impact of overcoming these challenges will lead to more accessible, efficient, and equitable healthcare systems, improving patient outcomes and reducing the burden on traditional healthcare infrastructures.

9.7 Conclusion

A fundamental framework for healthcare 5.0 modelling was proposed in this research. The advent of Healthcare 5.0 signifies a revolution in the provision of healthcare, providing previously unheard-of chances to improve patient care by using cutting-edge technologies like blockchain, big data analytics, artificial intelligence, and the Internet of Things (IoT). In the area of telemedicine, where these technologies have the potential to greatly optimize workflows, boost service delivery effectiveness, and offer more individualized, secure, and easily accessible healthcare solutions, this paradigm change is especially crucial. Healthcare 5.0 presents a number of issues for telemedicine, as we have discussed in this chapter, including patient engagement, interoperability, and data protection. Nevertheless, these obstacles can be overcome and the efficacy and sustainability of telemedicine systems increased by creative

Table 9.1 Challenges and solutions in telemedicine

Challenges	Description	Possible solutions	Impact
Data security and privacy	Ensuring the protection of patient data and compliance with regulations like HIPAA	Implement end-to-end encryption, multi-factor authentication, and secure cloud storage	Improved trust, reduced risk of data breaches, and compliance with regulations
Quality of care	Concerns that remote consultations might not match the quality of in person visits	Use high-quality video conferencing tools, AI assisted diagnostics, and remote monitoring systems	Enhanced care quality, ensuring remote consultations are as effective as in-person visits
Patient-provider relationship	Challenges in building trust and effective communication in virtual settings	Incorporatevirtual rapport-building techniques, invest in user friendly platforms, and providecontinuous follow-ups	Stronger patient provider relationships, improved adherence to treatment, and patient satisfaction
Technological reliability	Risk of disruptions due to poor internet connection or platform failures	Invest in redundant systems, cloud-based solutions, and reliable internet infrastructure	Increased reliability of telemedicine services, reduced service disruptions, and higher user trust
Clinical validation and standardization	Need for rigorous clinical validation and standardization of telemedicine practices	Conduct extensive clinical trials for telemedicine models and standardize protocols across platforms	Improved reliability, consistency, and trust in telemedicine technologies and practices
Access to technology	Limited access to devices or internet, creating disparities in healthcare access	Provide subsidies or lowcost devices, expand internet infrastructure, or offer offline solutions	Increased access to telemedicine services, especially in underserved areas, promoting equity
Training and adoption	Resistance to change and lack of digital tool familiarity among healthcare professionals	Provide ongoing training programs, offer incentives for adoption, and create user-friendly platforms	Faster adoption of telemedicine, improved efficiency, and confidence among healthcare providers

solutions that make use of these technologies' potential. The establishment of a strong ecosystem that enables the smooth integration of these technologies into telemedicine practices requires the cooperation of healthcare providers, technology developers, and legislators.

Furthermore, case studies in Healthcare 5.0 telemedicine show how these developments have practical applications. These case studies offer important insights into the real-world applications of telemedicine and the observable advantages it may offer to both patients and healthcare practitioners by highlighting successful deployments. These illustrations show how Healthcare 5.0 can help with urgent healthcare issues like improving diagnostic precision, lowering healthcare expenses, and expanding access to care in remote locations.

The future of healthcare is being shaped by emerging telemedicine trends like the usage of blockchain for secure data exchange, IoT devices for remote patient monitoring, and AI-driven diagnostics. These developments suggest a more integrated, data-driven healthcare system in which telemedicine is essential to delivering prompt, high-quality care. With the potential to build a more robust and flexible healthcare system, telemedicine's future in Healthcare 5.0 is incredibly promising.

In summary, by adopting the ideas of Healthcare 5.0, telemedicine can become more sustainable, effective, and accessible, which will enhance patient outcomes and make the healthcare system more successful. To fully realize the potential of telemedicine and ensure its seamless integration into the larger healthcare landscape, policymakers, technology developers, and healthcare professionals must continue to collaborate. Healthcare 5.0's transformational impact is further demonstrated by the case studies and emerging trends covered in this chapter, which also offer a glimpse into how healthcare will be delivered in an increasingly digital society.

References

1. M. D. Lytras, P. Papadopoulou, and A. Sarirete, "Smart Healthcare: emerging technologies, best practices, and sustainable policies," in *Innovation in Health Informatics*, Elsevier, 2020, pp. 3–38.
2. E. Mbunge, B. Muchemwa, S. Jiyane, and J. Batani, "Sensors and healthcare 5.0: transformative shift in virtual care through emerging digital health technologies," *Glob. Health J.*, vol. 5, no. 4, pp. 169–177, 2021.
3. A. Haleem, M. Javaid, R. P. Singh, R. Suman, and S. Rab, "Blockchain technology applications in healthcare: An overview," *International Journal of Intelligent Networks*, vol. 2, pp. 130–139, 2021.
4. A. Haleem and M. Javaid, "Industry 5.0 and its expected applications in medical field," *Curr. Med. Res. Pr.*, vol. 9, no. 4, pp. 167–169, 2019.
5. L. Gomathi, A. K. Mishra, and A. K. Tyagi, "Industry 5.0 for healthcare 5.0: Opportunities, challenges and future research possibilities," in *2023 7th International Conference on Trends in Electronics and Informatics (ICOEI)*, IEEE, 2023, pp. 204–213.
6. B. Mohanta, P. Das, and S. Patnaik, "Healthcare 5.0: A paradigm shift in digital healthcare system using artificial intelligence, IOT and 5G communication," in *2019 International Conference on Applied Machine Learning (ICAML)*, 2019.

7. M. Bhavin, S. Tanwar, N. Sharma, S. Tyagi, and N. Kumar, "Blockchain and quantum blind signature-based hybrid scheme for healthcare 5.0 applications," *J. Inf. Secur. Appl.*, vol. 56, no. 102673, p. 102673, 2021.
8. A. Rehman, S. Abbas, M. A. Khan, T. M. Ghazal, K. M. Adnan, and A. Mosavi, "A secure healthcare 5.0 system based on blockchain technology entangled with federated learning technique," *Comput. Biol. Med.*, vol. 150, no. 106019, p. 106019, 2022.
9. R. G. Maroju, S. G. Choudhari, M. K. Shaikh, S. K. Borkar, and H. Mendhe, "Role of telemedicine and digital technology in public health in India: A narrative review," *Cureus*, vol. 15, no. 3, p. e35986, 2023.
10. Y. Ma *et al.*, "Telemedicine application in patients with chronic disease: a systematic review and meta-analysis," *BMC Med. Inform. Decis. Mak.*, vol. 22, no. 1, p. 105, 2022.
11. A. Kumar, "The transformation of the Indian healthcare system," *Cureus*, vol. 15, no. 5, p. e39079, 2023.
12. H. R. Chi, M. de Fátima Domingues, H. Zhu, C. Li, K. Kojima, and A. Radwan, "Healthcare 5.0: In the perspective of consumer internet-of-things-based fog/cloud computing," *IEEE Trans. Consum. Electron.*, vol. 69, no. 4, pp. 745–755, 2023.
13. K. A. B. Ahmad, H. Khujamatov, N. Akhmedov, M. Y. Bajuri, M. N. Ahmad, and A. Ahmadian, "Emerging trends and evolutions for smart city healthcare systems," *Sustain. Cities Soc.*, vol. 80, no. 103695, p. 103695, 2022.
14. J. Li and P. Carayon, "Health Care 4.0: A vision for smart and connected Health Care," *IISE Trans. Healthc. Syst. Eng.*, vol. 11, no. 3, pp. 171–180, 2021.
15. A. Gupta and A. Singh, "Healthcare 4.0: recent advancements and futuristic research directions," *Wirel. Pers. Commun.*, vol. 129, no. 2, pp. 933–952, 2023.
16. D. Saraswat *et al.*, "Explainable AI for Healthcare 5.0: Opportunities and Challenges," *IEEE Access*, vol. 10, pp. 84486–84517, 2022.
17. N. Tsafack *et al.*, "A new chaotic map with dynamic analysis and encryption application in internet of health things," *IEEE Access*, vol. 8, pp. 137731–137744, 2020.
18. R. Pratap Singh, M. Javaid, A. Haleem, R. Vaishya, and S. Ali, "Internet of Medical Things (IoMT) for orthopaedic in COVID-19 pandemic: Roles, challenges, and applications," *J. Clin. Orthop. Trauma*, vol. 11, no. 4, pp. 713–717, 2020.
19. S. Swayamsiddha and C. Mohanty, "Application of cognitive Internet of Medical Things for COVID-19 pandemic," *Diabetes Metab. Syndr.*, vol. 14, no. 5, pp. 911–915, 2020.
20. J. U. Knickerbocker *et al.*, "Heterogeneous integration technology demonstrations for future healthcare, IoT, and AI computing solutions," in *2018 IEEE 68th Electronic Components and Technology Conference (ECTC)*, 2018.
21. I. Radanović and R. Likić, "Opportunities for use of blockchain technology in medicine," *Appl. Health Econ. Health Policy*, vol. 16, no. 5, pp. 583–590, 2018.
22. P. Zhang, J. White, D. C. Schmidt, G. Lenz, and S. T. Rosenbloom, "FHIRChain: Applying blockchain to securely and scalably share clinical data," *Computers and Biomedical Research*, vol. 95, pp. 195–208, 2018.
23. A. Tandon, A. Dhir, N. Islam, and M. Mäntymäki, "Blockchain in healthcare: A systematic literature review, synthesizing framework, and future research agenda," *Computers in Human Behavior*, vol. 117, 2020.
24. M. Wazid, A. K. Das, N. Mohd, and Y. Park, "Healthcare 5.0 security framework: Applications, issues and future research directions," *IEEE Access*, vol. 10, pp. 129429–129442, 2022.
25. M. Alshareef *et al.*, "Utilization of telemedicine in rhinologic practice during COVID-19 pandemic," *Am. J. Otolaryngol.*, vol. 42, no. 3, p. 102929, 2021.
26. K. Kędzierski, J. Radziejewska, A. Sławuta, M. Wawrzyńska, and J. Arkowski, "Telemedicine in cardiology: Modern technologies to improve cardiovascular patients' outcomes-A narrative review," *Medicina (Kaunas)*, vol. 58, no. 2, p. 210, 2022.
27. Punitha and Preetha, "Unleashing potential: a deep dive into AI-blockchain integration for UAV-enhanced tele-surgery," *Cogent Eng.*, vol. 11, no. 1, 2024.
28. P. Bhattacharya, M. S. Obaidat, D. Savaliya, S. Sanghavi, S. Tanwar, and B. Sadaun, "Metaverse assisted Telesurgery in Healthcare 5.0: An interplay of Blockchain and Explainable AI," in *2022*

International Conference on Computer, Information and Telecommunication Systems (CITS), 2022.

29. R. N. Hansen *et al.*, "Opportunities and barriers to rural telerobotic surgical health care in 2021: Report and research agenda from a stakeholder workshop," *Telemed. J. E. Health.*, vol. 28, no. 7, pp. 1050–1057, 2022.
30. D. B. Olawade, O. Z. Wada, A. Odetayo, A. C. David-Olawade, F. Asaolu, and J. Eberhardt, "Enhancing mental health with Artificial Intelligence: Current trends and future prospects," *Journal of Medicine, Surgery, and Public Health*, vol. 3, no. 100099, p. 100099, 2024.
31. A. Thakkar, A. Gupta, and A. De Sousa, "Artificial intelligence in positive mental health: a narrative review," *Front. Digit. Health*, vol. 6, p. 1280235, 2024.
32. S. Graham *et al.*, "Artificial intelligence for mental health and mental illnesses: An overview," *Curr. Psychiatry Rep.*, vol. 21, no. 11, p. 116, 2019.
33. V. K. P. Vudathaneni, R. B. Lanke, M. C. Mudaliyar, K. V. Movva, L. Mounika Kalluri, and R. Boyapati, "The impact of telemedicine and remote patient monitoring on healthcare delivery: A comprehensive evaluation," *Cureus*, vol. 16, no. 3, p. e55534, 2024.
34. S. Sikander, P. Biswas, and P. Kulkarni, "Recent advancements in telemedicine: Surgical, diagnostic and consultation devices," *Biomed. Eng. Adv.*, vol. 6, no. 100096, p. 100096, 2023.
35. J. Basulo-Ribeiro and L. Teixeira, "The future of healthcare with Industry 5.0: Preliminary interview-based qualitative analysis," *Future Internet*, vol. 16, no. 3, p. 68, 2024.
36. N. Agarwal, P. Jain, R. Pathak, and R. Gupta, "Telemedicine in India: A tool for transforming health care in the era of COVID-19 pandemic," *J. Educ. Health Promot.*, vol. 9, no. 1, p. 190, 2020.

Chapter 10
Intelligent Applications and Its Impact in Improving Brain-Health

Shivani Chauhan and Arvind K. Sharma

Abstract This chapter investigates the efficacy of artificial intelligence (AI)-enhanced mobile health (mHealth) applications in mitigating cognitive decline and promoting the mental well-being of individuals affected by neurocognitive disorders, as well as supporting their caregivers. Given the increasing global prevalence of cognitive impairment—characterized by progressive deterioration in memory, reasoning, and functional abilities—there is a pressing need for innovative interventions aimed at preserving cognitive functions, enhancing quality of life, and reducing caregiver burden. A comprehensive literature review was undertaken, synthesizing findings from 104 peer-reviewed studies focusing on AI-driven mHealth technologies. The review critically evaluates the impacts of these applications on core cognitive domains, including memory, executive functioning, language abilities, spatial and temporal orientation, and judgment in individuals experiencing cognitive deficits. Additionally, attention is given to design considerations, usability factors, and privacy protections, particularly in relation to aging populations, alongside an assessment of the extent to which caregiver support features are embedded within these technologies. The analysis reveals that, although many existing mHealth applications prioritize early detection, health monitoring, and predictive analytics, relatively few demonstrate efficacy in strengthening cognitive capacities such as memory retention, language processing, learning ability, and decision-making—areas notably vulnerable to neurodegenerative conditions. Furthermore, there remains a significant gap in the incorporation of caregiver support functionalities, indicating a broader systemic disconnect in addressing the multidimensional needs associated with cognitive care. The findings highlight a critical deficiency in the availability of integrated solutions that leverage AI, machine learning (ML), deep learning (DL), and Internet of Things (IoT) technologies within a unified platform capable of simultaneously addressing the needs of individuals with cognitive impairment and their caregivers.

S. Chauhan (✉) · A. K. Sharma
Yogananda School of Artificial Intelligence, Computers and Data Sciences, Shoolini University, Solan, Himachal Pradesh, India
e-mail: shivanichauhan09@shooliniuniversity.com

A. K. Sharma
e-mail: sharmaarvind0786@hotmail.com

© The Author(s), under exclusive license to Springer Nature Switzerland AG 2026
K. Dalip et al. (eds.), *Healthcare 5.0*, https://doi.org/10.1007/978-3-032-09582-4_10

This review underscores the considerable potential for the development of comprehensive mHealth ecosystems, wherein AI and IoT converge to facilitate sustained cognitive support for patients and real-time assistance for caregivers. A holistic, ethically grounded, and user-centered approach to mHealth innovation could significantly enhance accessibility, ensure robust privacy protections, and improve usability across diverse populations. Such advancements have the capacity to transform the paradigm of cognitive care by fostering more inclusive, effective, and compassionate strategies for managing neurocognitive health across the lifespan.

Keywords Dementia · Alzheimer disease · Artificial intelligence (AI) · Machine learning (ML) · Deep learning (DL) · Internet of things (IoT) · Mhealth applications · Cognitive function · Caregiver support · Elderly mental health · Privacy · Security

10.1 Introduction

As the global population continues to age, mental health challenges among older adults are becoming increasingly prevalent. Among these challenges, neurocognitive disorders such as dementia represent a significant and growing concern. Dementia is a syndrome resulting from various diseases that, over time, destroy nerve cells and cause structural damage to the brain, typically leading to a progressive decline in cognitive functions [1]. Consider, for instance, a grandparent struggling to recall a familiar name or becoming disoriented on their way home—such occurrences are unfortunately common. The World Health Organization (WHO) estimates that approximately 50 million people worldwide are currently living with dementia, with nearly 10 million new cases diagnosed each year; this figure is projected to rise to over 131.5 million by 2050 [1]. Dementia, which impairs memory, thinking, behavior, and the ability to perform everyday activities, has emerged as a leading cause of disability and dependency among the elderly [1]. These alarming statistics underscore the urgent need for effective and sustainable interventions to support both individuals living with cognitive decline and the caregivers who assist them. This chapter explores the potential for innovative technological solutions to bridge existing care gaps, with particular emphasis on the integration of artificial intelligence (AI), machine learning (ML), and Internet of Things (IoT) technologies. The Fig. 10.1. display the global prevalence of dementia. These advancements hold significant promise in enhancing cognitive functions, optimizing caregiver support, and ultimately improving the quality of life for individuals affected by dementia.

Caring for someone with dementia is often an arduous and highly stressful responsibility. Imagine a daughter managing her mother's cognitive training, social activities, and medication regimens while simultaneously balancing her own family and work obligations. While traditional caregiving methods have their merits, they often fall short in addressing the complex, multifaceted needs of both dementia patients and their caregivers. This is where innovative solutions, such as AI and machine learning

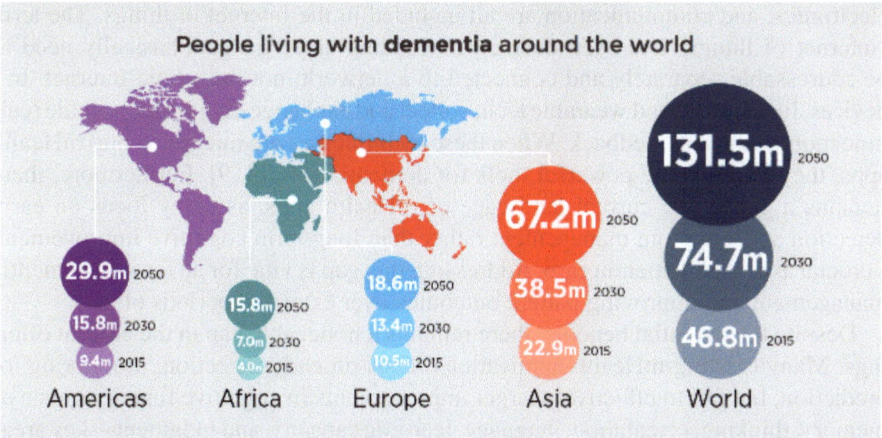

Fig. 10.1 Global prevalence of dementia

(ML)-enhanced mHealth applications, offer considerable promise. These advanced technologies hold the potential to provide personalized care, enhance communication, and facilitate more effective management of dementia care, despite the absence of a current cure [1]. The unique contribution of this review lies in its emphasis on long-term cognitive function improvement through the integration of AI, ML, and the Internet of Things (IoT)—an area that remains underexplored in existing literature. By evaluating these technologies, we aim to advance the ongoing dialogue around transforming dementia care from mere monitoring to actively enhancing cognitive abilities. Artificial intelligence (AI), machine learning (ML), and IoT are all reshaping the healthcare landscape. AI, often conceptualized as a "digital brain," performs tasks traditionally requiring human intelligence, such as learning, reasoning, and problem-solving. As a field of computer science, AI focuses on developing methods and software that enable machines to perceive their environment and use learning and intelligence to optimize decision-making processes and achieve specific goals [2].

Machine learning (ML), a subset of AI, involves systems capable of learning from data and improving over time without being explicitly programmed. This discipline centers around the development of statistical algorithms that can analyze data, generalize findings to new datasets, and perform tasks without pre-defined instructions. Recently, artificial neural networks have surpassed many earlier methodologies in terms of performance [3]. ML applications are widespread, spanning industries such as computer vision, speech recognition, email filtering, natural language processing, and healthcare [4, 5]. Predictive analytics is the term for it when used to solve business challenges. While not all machine learning is grounded in statistics, computational statistics is a major source of methodologies used in the discipline [6, 7]. Devices containing sensors, processing power, software, and other technologies that connect to other devices and systems over the Internet or other communications networks to exchange data are referred to as Internet of things, or IoT devices. Computer science, electronics, and communication are all included in the Internet of things. The term "internet of things" has been deemed misleading since gadgets basically need to be addressable separately and connected to a network, not the whole Internet. IoT devices, like sensors and wearable tech, collect and exchange data which provide real-time monitoring and feedback. When these technologies are integrated into mHealth apps, they can become powerful tools for dementia care [8, 9]. Furthermore, there remains a gap in the current landscape of mHealth apps, as many focus on early detection and symptom management rather than long-term cognitive improvement, a crucial aspect of dementia care. Addressing this gap is vital for advancing dementia management and improving patient outcomes over extended periods of use.

Despite the potential benefits, there remains a noticeable gap in the current offerings. Many existing mHealth applications focus on early detection, monitoring, or prediction, but fail to effectively target improvements in cognitive functions such as memory, thinking, orientation, language, learning capacity, and judgment—key areas affected by cognitive decline over time, which are crucial for providing meaningful feedback. This gap highlights the need for comprehensive solutions that leverage AI, ML, DL, and IoT technologies to address the multifaceted challenges associated with neurocognitive disorders. Integrating these advanced features and technologies into mHealth apps could substantially enhance the overall well-being of individuals experiencing cognitive impairment and their caregivers.

10.2 Background Literature

This chapter explores the role of AI-driven mobile health (mHealth) applications in addressing cognitive deterioration and supporting mental well-being in individuals affected by neurocognitive disorders, as well as their caregivers. As the global prevalence of memory-related conditions rises, the demand for effective, accessible interventions becomes increasingly urgent. Through an extensive review of 104 peer-reviewed studies, this chapter evaluates the effectiveness of AI-enhanced mHealth tools in mitigating symptoms of cognitive dysfunction—including memory loss, impaired judgment, disorientation, and language difficulties. It also analyzes critical factors such as usability, design, and data privacy, especially in the context of older users, while examining how well these technologies support caregivers. Although many digital tools emphasize early diagnosis and monitoring through predictive analytics, few directly enhance core cognitive abilities most impacted by neurodegenerative disorders. Moreover, there is a noticeable lack of integrated caregiver support features, revealing a fragmented approach in current solutions. The findings underscore a significant gap in the development of comprehensive platforms that merge AI, machine learning (ML), deep learning (DL), and Internet of Things (IoT) technologies. A more holistic, unified mHealth solution could offer continuous cognitive engagement for patients and real-time assistance for caregivers—while also addressing key concerns like accessibility, ease of use, and privacy. Such innovation has the potential to transform cognitive health management, offering a more compassionate and effective model for supporting those living with age-related cognitive decline and their caregivers.

10.2.1 Literature Analysis

This section will critically examine how AI technologies integrated with mHealth apps and also enhanced by IoT capabilities are advancing dementia care for the elderly. The analysis will also explore the role of these technologies in adapting cognitive interventions, predicting cognitive decline, and facilitating monitoring, while also addressing limitations related to long-term user engagement and consistent effectiveness across different cognitive functions.

Introduction to Dementia and the Need for Technological Interventions

Dementia is a global health challenge, affecting over 50 million people worldwide, with nearly 10 million new cases diagnosed each year [1]. By 2030, one in six people globally will be aged 60 years or older, and the number of individuals aged 80 years or older is expected to triple by 2050 [1]. This rapid aging of the population is accompanied by a rise in dementia cases, placing a significant burden on healthcare systems, families, and caregivers. In Malaysia, for instance, the prevalence of dementia among individuals aged 60 years and older is 8.5%, with higher rates observed among women, those with no formal education, and individuals living in rural areas [10]. These statistics underscore the urgent need for effective interventions to address the growing dementia crisis. The impact of dementia extends beyond patients to their caregivers, who often experience significant physical, emotional, and financial strain. In Malaysia, caregivers of people living with dementia (PLwD) report a lower quality of life (QoL) compared to caregivers of older adults without dementia. Factors such as the functional capacity of PLwD and the level of social support available to caregivers play a critical role in determining caregiver QoL [10]. Similarly, the World Health Organization (WHO) highlights that loneliness, social isolation, and elder abuse are key risk factors for mental health conditions among older adults, further exacerbating the challenges faced by this demographic [1]. Traditional care methods for dementia often fall short in addressing the complex and multifaceted needs of individuals with dementia and their caregivers [11]. These limitations are particularly evident in three key areas:

- Self-Management and Social Participation: Traditional approaches frequently lack the tools and training programs necessary to support individuals with dementia in maintaining self-management and engaging in meaningful activities, including social participation. For example, the FindMyApps program, a web-based selection tool combined with errorless learning training, was developed to address this gap. A pilot randomized controlled trial (RCT) found that while the program did not yield statistically significant improvements, small to moderate effect sizes suggested potential benefits for self-management and social participation, particularly among individuals over 70 and those with higher education levels [12].

- Safety and Wandering Risks: Assistive technology (AT) is often underutilized or inappropriately applied in addressing the safety and wandering risks of individuals with dementia living at home. A study examining AT installations found that caregiver support and living situation significantly influenced the type and frequency of AT interventions, rather than the level of risk or cognitive impairment [13]. This highlights the need for more person-centered approaches to AT selection and implementation.
- Medication Management: Traditional medication management for home-dwelling individuals with dementia is often fragmented and insufficient. The LIVE@Home.Path trial, a multicomponent cluster RCT, investigated the impact of medication reviews by general practitioners (GPs) on psychotropic drug use and behavioral and psychological symptoms of dementia (BPSD). While the trial found that medication reviews improved patient-GP communication, they did not significantly reduce psychotropic drug use or BPSD [14]. This underscores the complexity of pharmacologic treatment for this population and the need for broader, multicomponent interventions.

In light of the limitations of traditional care methods, there is growing interest in leveraging advanced technologies such as Artificial Intelligence (AI), Machine Learning (ML), the Internet of Things (IoT), and mobile health (mHealth) applications to address the gaps in dementia care. These technologies offer innovative solutions for early detection, personalized care, real-time monitoring, and caregiver support. For example:

- AI and ML: Algorithms can analyze large datasets to predict cognitive decline, personalize care plans, and optimize treatment strategies [15–21].
- IoT: Wearable sensors and smart home systems enable real-time monitoring of individuals with dementia, enhancing safety and reducing caregiver burden [22–25].
- mHealth Apps: Accessible platforms for cognitive training, remote monitoring, and caregiver support have shown promise in improving cognitive functions and facilitating communication between patients, caregivers, and healthcare providers [26, 27].

These technologies [28] hold immense potential to revolutionize dementia care by addressing the limitations of traditional methods and providing scalable, cost-effective solutions. However, their successful implementation requires addressing challenges related to usability, privacy, and accessibility, as well as ensuring equitable access across diverse populations.

Overview of AI, ML, and IoT Technologies in Dementia Care

Artificial Intelligence (AI), Machine Learning (ML), and the Internet of Things (IoT) are transformative technologies reshaping dementia care. AI refers to the simulation of human intelligence in machines, enabling tasks like learning, reasoning, and problem-solving, while ML, a subset of AI, involves algorithms that learn from data to improve performance over time [29, 30]. In dementia care, AI and ML are primarily used for early detection and diagnosis, analyzing large datasets such as electronic health records (EHRs) and biomarker data to predict cognitive decline with higher accuracy than traditional methods [15, 16, 31, 32]. Additionally, AI-powered systems support personalized treatment plans by tailoring interventions to individual patient needs, such as cognitive training programs or medication adjustments [16, 33–35]. However, challenges like data availability, algorithmic bias, and high costs hinder widespread adoption, necessitating further research to enhance reliability and transparency [15, 16]. The Internet of Things (IoT), a network of interconnected devices that collect and exchange data in real time, plays a critical role in real-time monitoring and enhancing safety for individuals with dementia. IoT devices, such as wearable sensors and smart home systems, track activities of daily living (ADLs), sleep patterns, and vital signs, enabling early intervention and reducing caregiver burden [22, 23, 27, 36, 37]. For example, IoT-enabled systems can detect wandering behavior or falls, triggering alerts for timely responses [23]. Despite their potential, IoT solutions face challenges related to user acceptance, cultural biases, and the need for user-centered design [38, 39], particularly for individuals with varying stages of dementia [22, 23]. Together, AI, ML, and IoT create a powerful ecosystem for dementia care, offering innovative solutions to improve patient outcomes and support caregivers, but their successful implementation requires addressing technical, ethical, and usability challenges [40, 41].

MHealth Applications for Dementia Care

Mobile health (mHealth) applications have emerged as valuable tools for supporting individuals with dementia and their caregivers, offering innovative solutions for cognitive training, remote monitoring, and lifestyle management [42–45]. A systematic search of mHealth apps for dementia prevention and risk factors revealed that while 57.9% of apps addressed modifiable lifestyle behaviors linked to reducing dementia risk, many focused on only one behavior, such as mental stimulation, emotion regulation [46] and lacked evidence-based credibility or adherence to data privacy standards [47]. Similarly, a review of mHealth interventions for cognitive impairment found that while some apps improved outcomes like cognition, function, and quality of life, the majority lacked rigorous evaluation, with inconsistent health outcomes and low-quality evidence [48]. These findings highlight the potential of mHealth apps to support dementia care but also underscore significant gaps in their development, including the need for multi-behavioral interventions, stronger evidence bases, and better privacy protections. Despite these challenges, mHealth apps have shown promise in enhancing user engagement and addressing the unique needs of individuals with dementia. A survey of 99 dementia care apps found no significant differences in quality between English and Chinese apps, though apps developed during the COVID-19 pandemic scored higher in engagement, suggesting a growing focus on user-centered design [49]. Additionally, a scoping review on mHealth usability for older adults with Alzheimer's disease and related dementias (ADRD) identified 42 barriers, such as cognitive and perceptual impairments, and proposed 20 design suggestions, including simplifying information display and breaking tasks into manageable steps [50]. These insights emphasize the importance of collaboration between dementia experts, designers, and caregivers to develop evidence-based, user-friendly mHealth apps that address the multifaceted needs of individuals with dementia and their caregivers [51–56].

Research Gaps and Challenges in AI-Enhanced mHealth for Cognitive Disorders

Despite significant advances in AI-powered mobile health solutions, several critical gaps remain in effectively supporting individuals with cognitive impairment and their caregivers.

1. Fragmented Solutions: Most current mHealth tools target isolated aspects of neurocognitive care—such as memory exercises or symptom tracking—without offering an integrated platform. There is a clear need for comprehensive systems that combine cognitive training, behavioral monitoring, and caregiver support within a unified, dementia-focused application.
2. Short-Term Focus in Research: Existing studies largely emphasize short-term outcomes, lacking longitudinal evidence on the sustained effectiveness of these technologies. Long-term data is essential to assess their real impact on cognitive health, quality of life, and caregiver burden.
3. Limited Representation in Study Populations: Research often lacks diversity, with a predominant focus on high-income regions and homogenous demographic groups. Broader, more inclusive studies are necessary to ensure applicability across cultures, socioeconomic backgrounds, and various stages of cognitive decline.
4. Ethical and Privacy Challenges: While concerns about ethics and privacy are acknowledged, few practical frameworks exist to guide real-world implementation. Further investigation is needed into safeguarding user data, obtaining informed consent, and ensuring responsible AI use in sensitive healthcare settings.
5. Underutilized AI-IoT Integration: Though both AI and the Internet of Things have been explored individually, integrated solutions that harness their combined potential for real-time, predictive, and personalized care remain scarce. Research should focus on developing intelligent systems that adapt dynamically to patients' evolving cognitive and behavioral needs.

6. Insufficient Support for Caregivers: Many digital interventions prioritize patient needs, often overlooking the essential role of caregivers. There is a need for tools that offer evidence-based resources, stress management strategies, and improved communication pathways to enhance caregiver resilience and effectiveness.
7. Lack of Standardized Evaluation Metrics: Cognitive outcomes are inconsistently measured across studies, making it difficult to compare findings or determine efficacy. Establishing standardized benchmarks would enhance reliability and cross-study comparisons in evaluating mHealth interventions.
8. Scalability and Real-World Adoption: Most solutions remain confined to controlled environments or small-scale pilots. To make a meaningful impact, research must address the scalability of these technologies—ensuring they can be seamlessly integrated into healthcare systems, remain cost-effective, and are widely accepted by users.

10.3 Analysis and Discussion

This section offers a comprehensive examination of research focused on mobile health (mHealth) applications enhanced by artificial intelligence (AI), machine learning (ML), and Internet of Things (IoT) technologies. The primary emphasis is on evaluating their effectiveness in enhancing cognitive function in individuals with cognitive impairments and providing meaningful support to caregivers. The analysis is organized around key thematic areas: the scope of cognitive functions addressed, the level of AI/ML integration, the impact on caregiver support, and the overall alignment of these technologies with the study's core objectives. To illustrate data patterns and emerging trends, visual tools such as heatmaps and cross-tabulations are employed. The findings reveal both the promise and the current limitations of AI-powered mHealth solutions. While notable progress has been made in areas such as early detection and cognitive monitoring, most applications still fall short of delivering fully integrated, multi-functional platforms that cater to the complex needs of both patients and caregivers. By synthesizing the available evidence, this section identifies critical research gaps and offers strategic recommendations aimed at advancing the development of holistic, inclusive, and scalable digital interventions in the field of neurocognitive care.

10.4 Summary of Findings

In the analysis of studies on mHealth applications enhanced with AI, ML, or DL technologies aimed at improving cognitive functions for dementia patients and support for caregivers, several key insights were found. The summary highlights the range of cognitive functions targeted, the degree of AI/ML involvement, and the effectiveness evidence reported. Heatmap of AI/ML Involvement by Effectiveness Evidence is presented to illustrate the distribution of studies based on AI/ML involvement and their corresponding effectiveness. This visual shows that a majority of studies incorporating AI/ML are associated with partial or promising effectiveness in cognitive improvements or no evidence at all of effectiveness. These findings suggest that AI/ML technologies have shown more reliable effectiveness in cognitive interventions when used independently, while IoT-involved studies may benefit from further development to enhance their impact, it also shows that IoT is used for caregiver support and not necessarily for improved cognitive functions. Non-technology approaches displayed varied effectiveness, suggesting potential inconsistencies in traditional methods. The effectiveness evidence v/s AL/ML involvement is shown in Fig. 10.2. Future research could focus on refining IoT integration in AI/ML applications to improve effectiveness outcomes and achieve a more holistic cognitive and caregiver support solution.

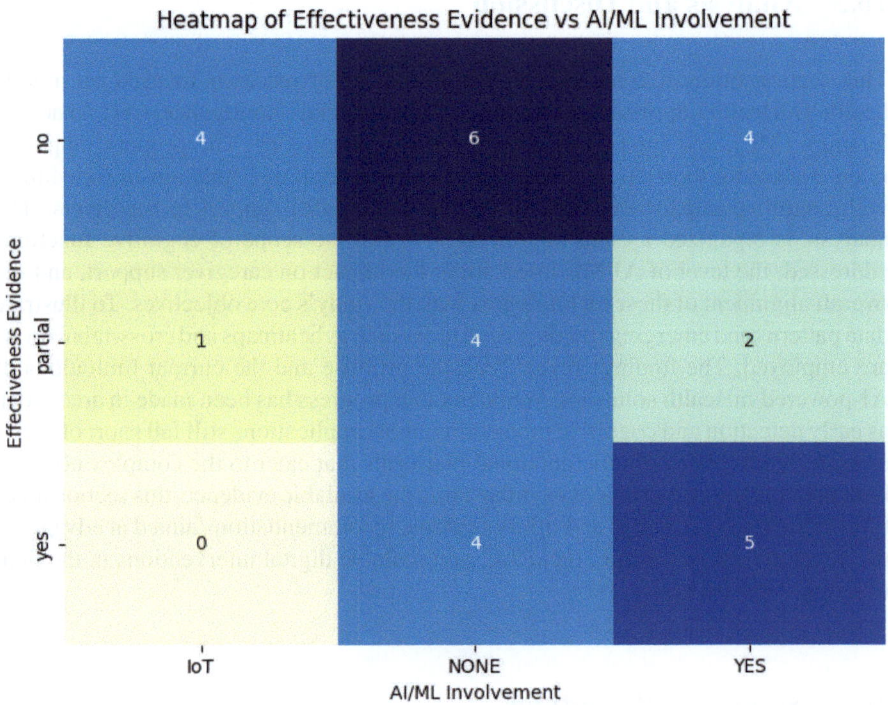

Fig. 10.2 Effectiveness evidence v/s AL/ML involvement

10.4.1 Impact on Caregivers

The analysis of caregiver-related findings reveals critical insights into the role of informal care providers in managing cognitive disorders and the extent to which current technologies—particularly mHealth applications—address their needs. Notably, there is a clear gap in research focusing on caregiver support, especially within the realm of mHealth solutions. Among the studies reviewed, 10 mHealth-related papers did not address caregiver impact at all, compared to only 3 in the category of other non-mHealth technologies. This indicates that mHealth apps are trailing behind in prioritizing caregiver considerations during development and evaluation. Although non-mHealth technologies appear slightly more responsive to caregiver needs, the overall number of studies dedicated to this aspect remains limited. Interestingly, a greater proportion of studies within the non-mHealth category reported positive outcomes for caregivers, suggesting that these tools—ranging from traditional monitoring devices to external support systems—may offer more immediate or practical benefits. These benefits often include real-time alerts, monitoring features, or structured caregiving tools that help reduce caregiver burden and enhance day-to-day management.

The lack of caregiver-focused features in mHealth applications points to a significant research and design gap. Many digital health solutions still fail to incorporate functionalities that directly assist caregivers, such as stress-reducing tools, communication platforms, educational resources, or task coordination systems. This oversight represents a missed opportunity to build more inclusive digital ecosystems that cater not just to individuals with cognitive impairments, but also to those who support them daily. Given that caregivers play a central role in the continuum of cognitive care—and frequently face emotional, physical, and psychological strain—technological interventions should be developed with a dual focus: enhancing patient outcomes while also alleviating caregiver challenges. The Fig. 10.3 shows the impact on caregivers support. Future mHealth innovations should aim to integrate comprehensive caregiver support features, ensuring that the technology holistically addresses the demands of dementia care.

The positive impact observed in other technologies might offer valuable lessons for the design and functionality of mHealth apps. Technologies that focus on direct caregiver support for instance through training modules, scheduling assistance, or monitoring tools could serve as an important benchmark for mHealth apps looking to expand their functionality. Further investigation into these technologies' success in supporting caregivers could provide insights that help refine mHealth applications to address caregiver needs more effectively. While mhealth apps have shown potential in improving patient outcomes, the relatively low number of studies focusing on caregiver impact points to a significant area for improvement. It is essential that future mHealth applications be developed with a dual focus on both patient and caregiver needs, integrating features that provide direct, actionable support to caregivers. Enhancing the caregiver experience could not only improve adherence to treatment

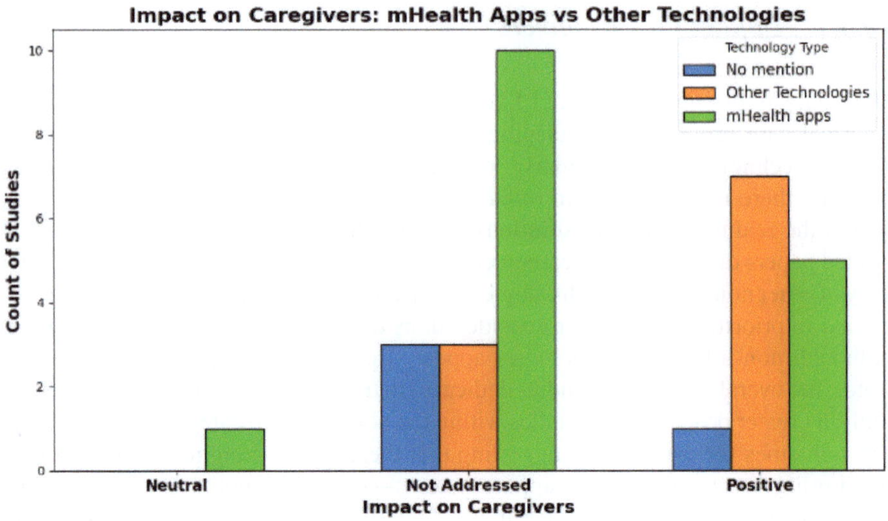

Fig. 10.3 Impact on caregivers support

protocols but also foster more positive care environments for patients, ultimately contributing to better health outcomes.

To ensure the app's effectiveness, it should integrate evidence-based cognitive interventions from existing research. VR-based cognitive training programs, as highlighted in some studies, can be adapted into the app as interactive modules, allowing patients to practice tasks like grocery shopping or navigating a virtual environment. Additionally, ICT-based interventions, such as gamified memory exercises and attention-building tasks, can be incorporated to improve global cognition and quality of life. By integrating these proven interventions, the app can provide a comprehensive cognitive enhancement solution that addresses multiple domains, including memory, judgment, and reasoning can further enhance cognitive functions by providing real-time adaptation and feedback. Machine learning algorithms can analyze patient performance during cognitive training exercises and dynamically adjust the difficulty level or type of task based on real-time data. For example, if a patient struggles with a memory task, the app can simplify the task or provide additional hints. Conversely, if a patient excels, the app can introduce more complex challenges to maintain engagement. Reinforcement learning algorithms can optimize interventions by continuously learning from patient responses and tailoring future tasks to maximize cognitive gains. The app can also leverage multimodal data from IoT devices and electronic health records (EHRs) to enhance cognitive interventions. Wearable sensors can track a patient's physical activity and sleep patterns, which are closely linked to cognitive health. AI algorithms can analyze this data to recommend personalized lifestyle changes, such as increased physical activity or improved sleep hygiene, to complement cognitive training. EHR data can provide insights into a patient's medical history and medication use, allowing the app to tailor interventions based on individual health conditions [23, 27].

Finally, the app should include a progress tracking feature that allows patients and caregivers to monitor cognitive improvements over time. AI algorithms can analyze performance data from cognitive training exercises and generate personalized insights, such as areas of strength and weakness. Progress reports can be shared with healthcare providers to inform clinical decision-making and adjust treatment plans as needed. Studies have shown that progress tracking and personalized feedback significantly improve patient engagement and outcomes, making this feature a critical component of the app. Developing an integrated AI/ML and IoT-powered mHealth app requires close collaboration between developers and healthcare providers to ensure that the solution is both technically robust and clinically effective. Developers bring expertise in AI algorithms, IoT integration, and user-centered design, while healthcare providers offer critical insights into patient needs, clinical workflows, and evidence-based interventions. Regular communication and co-design sessions can facilitate knowledge exchange, ensuring that the app's features align with real-world care requirements. For example, healthcare providers can help developers understand the specific cognitive domains that need to be targeted, such as memory, judgment, and reasoning, while developers can translate these needs into scalable, user-friendly features. Additionally, healthcare providers can assist in pilot testing the app in clinical settings, providing feedback on usability, effectiveness, and integration into existing care protocols. By working together, developers and healthcare providers can create a solution that not only leverages cutting-edge technology but also addresses the practical challenges of dementia care, ultimately improving outcomes for both patients and caregivers.

The proposed architecture of mobile application is displayed in Fig. 10.4. The proposed system is an intelligent, connected mHealth application created to assist individuals with dementia, along with their caregivers and healthcare providers. By integrating artificial intelligence, machine learning, and Internet of Things (IoT) technologies, it offers personalized care and real-time support. The app features three main user interfaces: patients engage with cognitive training games, daily mood check-ins, and gentle memory and routine reminders; caregivers receive real-time alerts, patient updates, and access to stress-reducing resources; and healthcare professionals utilize a dashboard that presents detailed reports, enabling them to adjust treatment plans and deliver timely interventions.

10.4.2 Future Research Directions

Given the complex and evolving nature of cognitive decline, along with the ambitious goal of integrating AI, ML, and DL technologies into IoT-enabled mobile health (mHealth) applications, future research must move beyond fragmented or purely technical approaches. The findings of this review highlight that meaningful progress in this field demands a holistic, ethically grounded, and person-centered strategy.

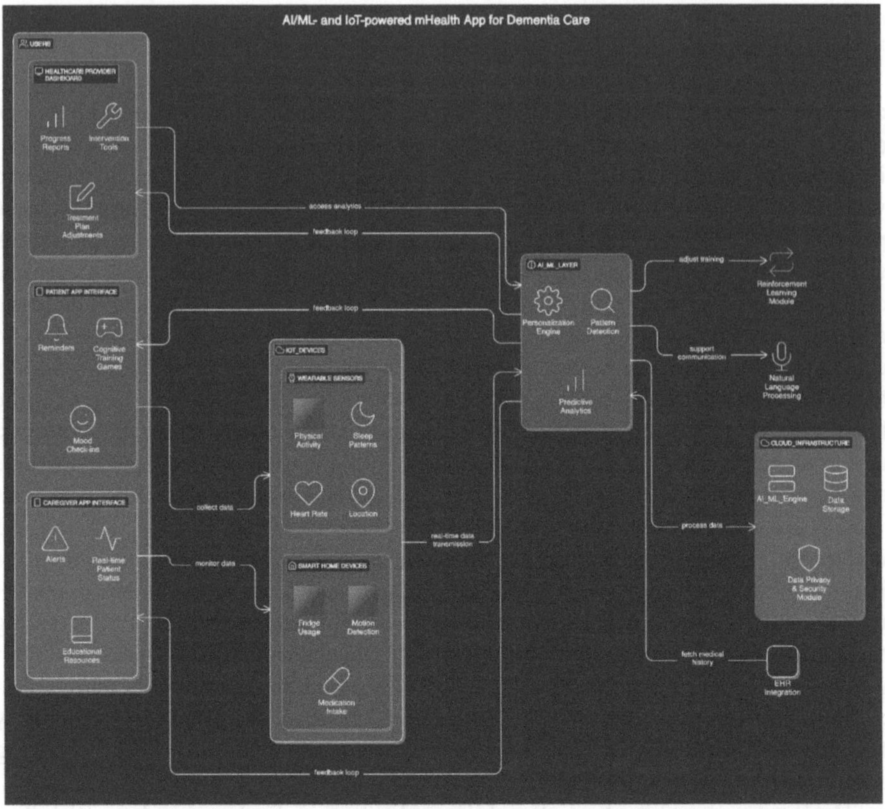

Fig. 10.4 Proposed architecture of mobile application

Technology alone is not the solution; its true value lies in how effectively it resonates with the real-world experiences of those it aims to support. AI-driven cognitive care must be crafted not just for efficiency and prediction but also for adaptability, empathy, and enduring assistance. It should respond to the shifting needs of both individuals and their caregivers, offering interventions that are as compassionate as they are intelligent. This review reinforces the idea that the future of cognitive health support will depend not solely on technological innovation but on our ability to embed these advancements within inclusive, sustainable, and human-centered frameworks. In light of this, we propose the following future research directions:

1. Longitudinal and Multisite Evaluations

   Most current studies show that people with dementia might improve a little in the short term when using certain apps or tools, but that's just the beginning. What we don't really know is how well these improvements last over time. That's a big gap. Future research needs to go deeper. We need long-term studies, not just a few weeks or months, but trials that stretch across one to three years. And these shouldn't happen in just one place or with one kind of group. We need diverse voices, people from different cultures, backgrounds, and levels of dementia. These studies should follow both patients and caregivers closely. They should ask: Does memory continue to improve? What about reasoning, judgment, learning, and language? But it's not just about numbers or test scores. It's about the real experience—the emotional stress, the day-to-day struggles, and even the unseen pressure caregivers carry over long periods. Only by painting this full picture can we truly understand how helpful these AI-powered, IoT-connected health apps can be in real life and not just in theory.

2. Adaptive and Explainable AI Models

   To truly build trust and make these tools feel personal, future apps should use adaptive AI smart systems that can change and grow with each user. People with dementia don't follow a straight path; their needs and behaviors can shift day by day, even hour by hour. So, the AI needs to keep up, adjusting in real-time based on what it sees, how the person is feeling, acting, or thinking.This means learning from real-world use: how someone interacts with the app, the data collected from sensors, and the patterns that emerge over time. But for people to actually feel comfortable using this kind of technology, it's not enough for it to just "work.", they need to understand it. That's where Explainable AI (XAI) comes in. Instead of being a black box, the system should be able to explain why it made a certain suggestion or decision especially when the stakes are high and mistakes could really affect someone's health or well-being.

3. Modular All-in-One mHealth Frameworks

   Right now, many of the tools built to support dementia care feel like scattered puzzle pieces—each one focusing on just a small part of the bigger picture. Some apps might help with cognitive games, others monitor daily activity, and a few provide basic alerts for caregivers. But for patients and caregivers dealing with the *complex, day-to-day realities* of dementia, juggling between separate tools can feel overwhelming and unsustainable. What's truly needed is a comprehensive, all-in-one solution—a flexible digital ecosystem that brings everything together under one roof. Imagine a platform that not only helps patients practice memory and language skills but also keeps track of their physical activity, vital signs, sleep quality, and mood, all through IoT devices. At the same time, it would alert caregivers in real time if something's off, offer personalized mental health check-ins, and use AI to predict changes before they happen. And this platform shouldn't be one-size-fits-all. It needs to be modular and customizable and able to adapt to different cultures, languages, care routines, and levels of cognitive impairment. A caregiver in rural India and one in urban France may need different things, but they both deserve a system that feels intuitive, secure, and deeply helpful. In the end, the goal is to reduce friction, build trust, and create a seamless experience where care feels connected, smart, and genuinely human-centered.

4. Ethical and Privacy-Respectful Design

   As artificial intelligence becomes more capable and complex, the ethical questions surrounding its use in dementia care grow just as urgent. It's not enough to build smart systems, we must build fair, respectful, and trustworthy ones. A key challenge lies in algorithmic bias: Many existing AI models are trained on data that underrepresent people from marginalized or minority communities. When these systems are then used in real-world healthcare settings, they risk reinforcing inequalities rather than solving them. To move forward, future research must prioritize inclusive datasets that reflect the full diversity of people living with dementia across race, gender, culture, language, socioeconomic background, and cognitive stages. At the same time, the algorithms themselves need to be transparent and interpretable so that patients, caregivers, and clinicians can understand how and why certain decisions are made. Privacy is another critical concern. Dementia patients often cannot fully advocate for themselves, so systems must be designed to protect sensitive health information without sacrificing personalization. Promising techniques like federated learning, where data stays on the device instead of being uploaded, and homomorphic encryption, which allows encrypted data to be used in computation, can help strike this delicate balance. Lastly, we need to rethink how consent works in digital health. Long, complex privacy agreements aren't practical or ethical for someone with cognitive impairment. Instead, consent should be simple, clear, and revisitable, ensuring that users (or their legal guardians) understand their rights and retain control over how their data is used at every stage of their journey.

5. Equity-Driven Deployment and Cost-Effectiveness Research

   Dementia often places its heaviest burden on people living in underserved or rural areas, where access to specialized healthcare is limited. Because of this, future research should focus on making mHealth apps more inclusive, accessible, and fair for everyone, no matter where they live. These apps need to be designed with real-world challenges in mind. For example, in regions where internet connectivity is poor or unstable, the apps should still work reliably. This means building them to run smoothly in low-bandwidth environments, with features that can function offline and without the need for expensive or advanced devices. In addition to technical accessibility, financial accessibility is just as important. If an app is too costly to use or maintain, its impact will be limited. That's why future studies should also look into how to make these solutions cost-effective in the long term. This could include comparing different pricing strategies, such as subscription models or one-time payments, as well as exploring partnerships between governments, private companies, and nonprofit organizations. Incentives from public policies might also help support wider adoption, especially in low-income areas.Ultimately, the goal is to create mHealth tools that are not just innovative, but practical and scalable. By focusing on affordability, low-resource compatibility, and thoughtful design, researchers can help ensure that dementia care powered by AI and mHealth technologies reaches the people who need it most.

6. Emerging Technologies: Digital Twins, 6G, and Beyond

   To truly transform dementia care in meaningful and lasting ways, future research must embrace the full spectrum of emerging technologies that are beginning to reshape healthcare. One powerful idea is the use of digital twins, virtual replicas of real patients that can mimic their behaviors, health patterns, and treatment responses. These digital twins could allow healthcare teams to run safe simulations, predict disease progression, and personalize care plans before trying anything in the real world. At the same time, the rise of next-generation networks like 6G holds immense promise. With its ultra-fast speeds and almost zero delays, 6G could make it possible to analyze massive streams of data in real time—everything from wearable biosensor readings to environmental inputs. This means care teams and caregivers could respond to even subtle changes in a patient's condition instantly, before small issues become serious ones. But the possibilities don't stop there. Neuro-symbolic AI, which combines the strengths of deep learning and symbolic reasoning, could help create smarter, more explainable systems that understand not just data patterns but also context and meaning. Context-aware computing, too, can make these systems more responsive to real-life situations, adjusting recommendations based on a person's environment, mood, or behavior. And let's not overlook immersive technologies like augmented and virtual reality. These tools could open new doors for cognitive rehabilitation, offering engaging, interactive exercises that not only support memory and attention but also provide joy and motivation through playful learning. Together, these innovations could enrich dementia care in ways that were previously unimaginable, turning complex data into insight, isolation into connection, and static care into a living, evolving system tailored to each individual's needs.

7. Feedback-Informed Development and Living Reviews

   At the heart of any meaningful innovation in dementia care lies a simple truth, mHealth apps must grow and change alongside the very people they are built for. The journey of dementia is not static, and neither should be the technology that supports it. Future systems should be designed to listen actively, evolve constantly, and respond intuitively. To achieve this, continuous feedback must become a core part of app design. Small yet powerful tools like voice diaries, mood check-ins, or quick reaction prompts can capture how a patient or caregiver is really feeling day to day, moment to moment. These inputs, when collected over time, offer rich stories far beyond numbers and checkboxes. By harnessing the power of Natural Language Processing (NLP), future research can make sense of this unstructured feedback, finding patterns, concerns, or even subtle cries for help hidden in everyday language. These insights shouldn't just vanish into storage, they should fuel real-time improvements in the app, guiding how features are adjusted, how notifications are sent, or how support is delivered. But the impact can go further. When collected across large populations, this kind of lived data can inform clinical research, shape care protocols, and update best practices for the whole field. It creates a loop, a living ecosystem, where the experiences of users directly shape the tools built for them. In the end, this is about more than technology. It's about empathy in design, intelligence in response, and humility in innovation. It's a reminder that the best systems don't just serve; they learn, they listen, and they grow with us.

In summary as shown in Fig. 10.5, the future of dementia care must move beyond isolated tech fixes and instead embrace a more connected, inclusive, and ethically aware approach. Research should no longer focus solely on technological advancement in silos, but rather blend disciplines medicine, AI, psychology, design, and public health—to create solutions that truly work in the messiness of real life. The true potential of AI, machine learning, and deep learning in mHealth apps lies not just in their impressive algorithms or predictive power but in their ability to serve as thoughtful, adaptive companions on the journey through dementia. When paired with IoT, these tools can offer more than monitoring; they can offer understanding, routine, reassurance, and even moments of joy. This vision requires us to center equity, empathy, and ethics at every stage, from design to deployment. It's about building systems that evolve with people, that listen and respond, and that support both those living with dementia and the loved ones who care for them. Only then can we realize the full promise of intelligent health technologies: not just as machines but as partners in preserving dignity, connection, and quality of life.

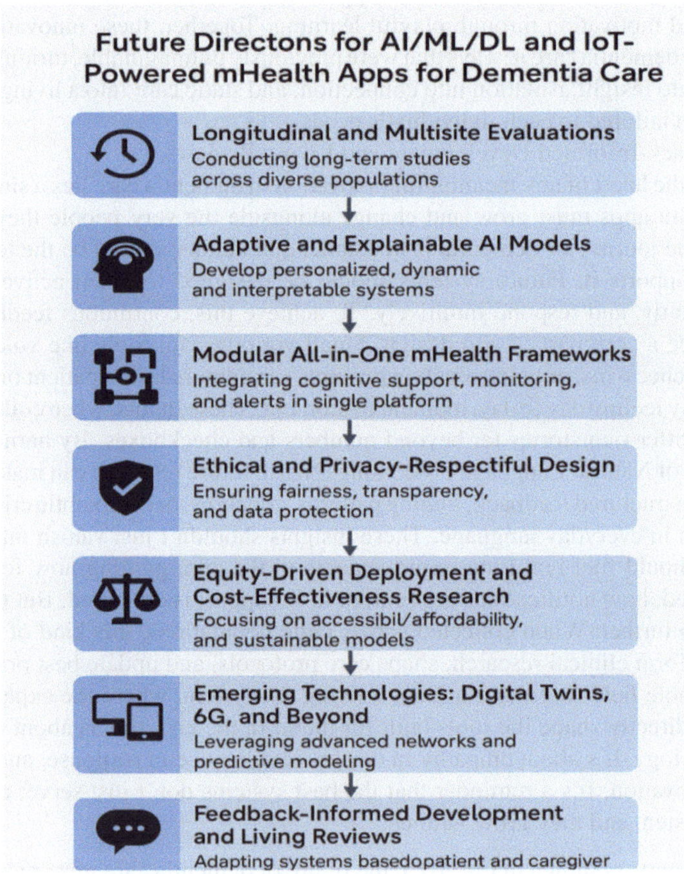

Fig. 10.5 Summary of the future directions for AI powered mHealth apps

10.5 Conclusion

This study set out with a hopeful vision: harnessing AI, machine learning, and IoT technologies to better support individuals experiencing cognitive decline and their caregivers. We've witnessed significant advancements—AI can now detect early indicators of cognitive impairment, and IoT systems enable caregivers to monitor loved ones remotely. mHealth applications are becoming more accessible, providing memory aids, medication reminders, and emotional well-being check-ins. Yet, despite this progress, a crucial element remains missing. Most existing solutions address only fragments of the broader challenge. Some platforms help caregivers track health data, while others offer cognitive training exercises for patients—but few integrate these functions into a cohesive whole. What's needed is a comprehensive, dedicated platform for cognitive care: one that seamlessly combines early detection, personalized interventions, real-time assistance, and user-friendly design. A platform that feels more like a true companion than just another technological tool. The future of digital cognitive care lies in developing such a solution—one that is affordable, ethically designed, adaptable, and deeply centered around human needs. It must serve people from all walks of life, regardless of income level, education, or geographical location. Emerging technologies like adaptive AI and digital twins should be leveraged not simply for their sophistication, but because they can deliver more personalized, meaningful, and responsive care. Privacy must be safeguarded without compromising ease of use.

The review paper does not present that ultimate solution—not yet. Instead, it lays the conceptual groundwork by mapping the current research landscape, spotlighting successful innovations, and identifying critical gaps. It serves as a foundation and a call to action for future developers, researchers, and healthcare professionals. We also emphasize the importance of real-world testing: longitudinal studies involving diverse populations, continuous feedback from both patients and caregivers, and strategies for scalability and affordability. Only through such comprehensive efforts can we truly assess the long-term impact. Ultimately, the aim is to create a transformative mHealth platform that is not only technologically advanced but also profoundly human-centered. By doing so, we move closer to a future where digital health tools genuinely enhance the quality of life, dignity, and well-being of individuals living with cognitive challenges and those who care for them.

References

1. World Health Organization, "Mental health of older adults," 2017.
2. R. Stuart and N. Peter, Artificial Intelligence: A Modern Approach. 2022.
3. K. B. Gomes et al., "Machine Learning-Based Routine Laboratory Tests Predict One-Year Cognitive and Functional Decline in a Population Aged 75+ Years.," Brain Sci, vol. 13, no. 4, Apr. 2023, https://doi.org/10.3390/brainsci13040690.
4. A. Javeed, A. L. Dallora, J. S. Berglund, A. Ali, L. Ali, and P. Anderberg, "Machine Learning for Dementia Prediction: A Systematic Review and Future Research Directions," J Med Syst, vol. 47, no. 1, Dec. 2023, https://doi.org/10.1007/s10916-023-01906-7.

5. M. S. Treder, S. Lee, and K. A. Tsvetanov, "Introduction to Large Language Models (LLMs) for dementia care and research.," Frontiers in dementia, vol. 3, p. 1385303, May 2024, https://doi.org/10.3389/frdem.2024.1385303.
6. J. Hu, H. Niu, J. Carrasco, B. Lennox, and F. Arvin, "Voronoi-Based Multi-Robot Autonomous Exploration in Unknown Environments via Deep Reinforcement Learning," IEEE Trans VehTechnol, vol. 69, no. 12, 2020, https://doi.org/10.1109/TVT.2020.3034800.
7. "Machine Learning - IBM Research." Accessed: Jan. 05, 2025. [Online]. Available: https://research.ibm.com/topics/machine-learning
8. A.Gazis, "What is IoT? The Internet of Things explained," Academia Letters, Jun. 2021, https://doi.org/10.20935/AL1003.
9. M. Shafiq, Z. Gu, O. Cheikhrouhou, W. Alhakami, and H. Hamam, "The Rise of 'internet of Things': Review and Open Research Issues Related to Detection and Prevention of IoT-Based Security Attacks," WirelCommun Mob Comput, vol. 2022, 2022, https://doi.org/10.1155/2022/8669348.
10. S. S. Ganapathy et al., "Prevalence of dementia and quality of life of caregivers of people living with dementia in Malaysia," GeriatrGerontolInt, vol. 20, no. S2, pp. 16–20, Dec. 2020, https://doi.org/10.1111/ggi.14031.
11. G. Livingston et al., "Dementia prevention, intervention, and care: 2020 report of the Lancet Commission," The Lancet, vol. 396, no. 10248, pp. 413–446, Aug. 2020, https://doi.org/10.1016/S0140-6736(20)30367-6.
12. K. M. Beentjes et al., "Impact of the FindMyApps program on people with mild cognitive impairment or dementia and their caregivers; an exploratory pilot randomised controlled trial," DisabilRehabil Assist Technol, vol. 18, no. 3, pp. 253–265, 2023, https://doi.org/10.1080/17483107.2020.1842918.
13. E. Curnow, R. Rush, S. Gorska, and K. Forsyth, "Differences in assistive technology installed for people with dementia living at home who have wandering and safety risks," BMC Geriatr, vol. 21, no. 1, Dec. 2021, https://doi.org/10.1186/s12877-021-02546-7.
14. M. H. Gedde et al., "The impact of medication reviews by general practitioners on psychotropic drug use and behavioral and psychological symptoms in home-dwelling people with dementia: results from the multicomponent cluster randomized controlled LIVE@Home.Path trial," BMC Med, vol. 20, no. 1, Dec. 2022, https://doi.org/10.1186/s12916-022-02382-5.
15. A.E. Andargoli, N. Ulapane, T. A. Nguyen, N. Shuakat, J. Zelcer, and N. Wickramasinghe, "Intelligent decision support systems for dementia care: A scoping review," ArtifIntell Med, vol. 150, Apr. 2024, https://doi.org/10.1016/J.ARTMED.2024.102815.
16. D. M. Lyall et al., "Artificial intelligence for dementia-Applied models and digital health," Alzheimers Dement, vol. 19, no. 12, pp. 5872–5884, Dec. 2023, https://doi.org/10.1002/ALZ.13391.
17. C.-C. Wu, C.-H. Su, M. M. Islam, and M.-H. Liao, "Artificial Intelligence in Dementia: A Bibliometric Study.," Diagnostics (Basel), vol. 13, no. 12, Jun. 2023, https://doi.org/10.3390/diagnostics13122109.
18. A,Merkin, R. Krishnamurthi, and O. N. Medvedev, "Machine learning, artificial intelligence and the prediction of dementia," CurrOpin Psychiatry, vol. 35, no. 2, pp. 123–129, Mar. 2022, https://doi.org/10.1097/YCO.0000000000000768.
19. F. Cheng et al., "Artificial intelligence and open science in discovery of disease-modifying medicines for Alzheimer's disease," Cell Rep Med, vol. 5, no. 2, Feb. 2024, https://doi.org/10.1016/j.xcrm.2023.101379.
20. S. A. Graham et al., "Artificial intelligence approaches to predicting and detecting cognitive decline in older adults: A conceptual review," Psychiatry Res, vol. 284, Feb. 2020, https://doi.org/10.1016/j.psychres.2019.112732.
21. A.W. Salehi, P. Baglat, and G. Gupta, "Alzheimer's Disease Diagnosis using Deep Learning Techniques," Int J EngAdvTechnol, vol. 9, no. 3, pp. 874–880, Feb. 2020, https://doi.org/10.35940/IJEAT.C5345.029320.
22. A.K. Behera, J. Condell, S. Dora, D. S. Gibson, and G. Leavey, "State-of-the-Art Sensors for Remote Care of People with Dementia during a Pandemic: A Systematic Review," Sensors (Basel), vol. 21, no. 14, Jul. 2021, https://doi.org/10.3390/S21144688.

23. M. Esquer-Rochin, L. F. Rodríguez, and J. O. Gutierrez-Garcia, "The Internet of Things in dementia: A systematic review," Internet of Things (Netherlands), vol. 22, Jul. 2023, https://doi.org/10.1016/J.IOT.2023.100824.
24. S. Ottaviani and F. Monacelli, "Rethinking Dementia Risk Prediction: A Critical Evaluation of a Multimodal Machine Learning Predictive Model," Journal of Alzheimer's Disease, vol. 97, no. 3, pp. 1097–1100, Jan. 2024, https://doi.org/10.3233/JAD-231071.
25. W. Salehi, G. Gupta, S. Bhatia, D. Koundal, A. Mashat, and A. Belay, "IoT-Based Wearable Devices for Patients Suffering from Alzheimer Disease," Contrast Media Mol Imaging, vol. 2022, 2022, https://doi.org/10.1155/2022/3224939.
26. P. Bhatt, J. Liu, Y. Gong, J. Wang, and Y. Guo, "Emerging Artificial Intelligence-Empowered mHealth: Scoping Review," JMIR MhealthUhealth, vol. 10, no. 6, Jun. 2022, https://doi.org/10.2196/35053.
27. A,Sheikhtaheri and F. Sabermahani, "Applications and Outcomes of Internet of Things for Patients with Alzheimer's Disease/Dementia: A Scoping Review," Biomed Res Int, vol. 2022, 2022, https://doi.org/10.1155/2022/6274185.
28. S. Shu and B. K. Woo, "Use of technology and social media in dementia care: Current and future directions," World J Psychiatry, vol. 11, no. 4, p. 109, Apr. 2021, https://doi.org/10.5498/WJP.V11.I4.109.
29. W. Qi et al., "Mapping Knowledge Landscapes and Emerging Trends in AI for Dementia Biomarkers: Bibliometric and Visualization Analysis," J Med Internet Res, vol. 26, 2024, https://doi.org/10.2196/57830.
30. S. FarabiMaleki, M. Yousefi, N. Sobhi, A. Jafarizadeh, R. Alizadehsani, and J. M. Gorriz-Saez, "Artificial Intelligence in Eye Movements Analysis for Alzheimer's Disease Early Diagnosis," Curr Alzheimer Res, vol. 21, no. 3, pp. 155–165, Jun. 2024, https://doi.org/10.2174/0115672050322607240529075641.
31. M. Bucholc et al., "Artificial intelligence for dementia research methods optimization.," Alzheimers Dement, vol. 19, no. 12, pp. 5934–5951, Dec. 2023, https://doi.org/10.1002/alz.13441.
32. A.B. Varghese, M. Gokilavani, M. Kunjachan, A. Namboodhiri, and G. Menezes, "AI Based Caregiver for Dementia Patients," Proceedings of the 5th International Conference on I-SMAC (IoT in Social, Mobile, Analytics and Cloud), I-SMAC 2021, pp. 601–605, 2021, https://doi.org/10.1109/I-SMAC52330.2021.9640970.
33. S. Lee-Cheong, S. Amanullah, and M. Jardine, "New assistive technologies in dementia and mild cognitive impairment care: A PubMed review," Asian J Psychiatr, vol. 73, Jul. 2022, https://doi.org/10.1016/J.AJP.2022.103135.
34. R. Gathercole et al., "Assistive technology and telecare to maintain independent living at home for people with dementia: The ATTILA RCT," Health Technol Assess (Rockv), vol. 25, no. 19, pp. 1–156, 2021, https://doi.org/10.3310/hta25190.
35. S. H. Pang, S. F. Lim, and C. J. Siah, "Online memory training intervention for early-stage dementia: A systematic review and meta-analysis," J AdvNurs, vol. 77, no. 3, pp. 1141–1154, Mar. 2021, https://doi.org/10.1111/JAN.14664.
36. M. Ianculescu, E.-A. Paraschiv, and A. Alexandru, "Addressing Mild Cognitive Impairment and Boosting Wellness for the Elderly through Personalized Remote Monitoring.," Healthcare (Basel), vol. 10, no. 7, Jun. 2022, https://doi.org/10.3390/healthcare10071214.
37. H. H. Kim and N. H. Jung, "Effects of Assistive Technology Application in Dementia Intervention for People with Mild Cognitive Impairment & Mild Alzheimer Type Dementia and Caregiver," AlternTher Health Med, vol. 29, no. 2, pp. 104–111, Mar. 2023.
38. A.vonBosse, A. Bejan, M. Wessel, and C. Kunze, "Design and evaluation options of multimedia applications to support the care of people with mild dementia," Z GerontolGeriatr, vol. 57, no. 5, pp. 382–388, Aug. 2024, https://doi.org/10.1007/s00391-023-02280-2.
39. S. Fox et al., "Co-design of a Smartphone App for People Living with Dementia by Applying Agile, Iterative Co-design Principles: Development and Usability Study," JMIR MhealthUhealth, vol. 10, no. 1, Jan. 2022, https://doi.org/10.2196/24483.

40. Q. Wang et al., "Usability evaluation of mHealth apps for elderly individuals: a scoping review," BMC Med Inform DecisMak, vol. 22, no. 1, Dec. 2022, https://doi.org/10.1186/S12911-022-02064-5.
41. J. A. Nkodo et al., "The Role of Telemedicine in the Management of the Behavioral and Psychological Symptoms of Dementia: A Systematic Review," American Journal of Geriatric Psychiatry, vol. 30, no. 10, pp. 1135–1150, Oct. 2022, https://doi.org/10.1016/J.JAGP.2022.01.013.
42. C. Göransson, Y. Wengström, M. Hälleberg-Nyman, A. Langius-Eklöf, K. Ziegert, and K. Blomberg, "An app for supporting older people receiving home care - usage, aspects of health and health literacy: a quasi-experimental study," BMC Med Inform DecisMak, vol. 20, no. 1, Sep. 2020, https://doi.org/10.1186/S12911-020-01246-3.
43. Y. H. Chou, C. Lin, S. H. Lee, Y. W. C. Chien, and L. C. Cheng, "Potential Mobile Health Applications for Improving the Mental Health of the Elderly: A Systematic Review," ClinInterv Aging, vol. 18, pp. 1523–1534, 2023, https://doi.org/10.2147/CIA.S410396.
44. Q.-X. Hong et al., "The effectiveness of virtual passport, an app-based intervention, for dementia care.," Front Psychiatry, vol. 15, p. 1457923, 2024, https://doi.org/10.3389/fpsyt.2024.1457923.
45. J. Krafft, B. Barisch-Fritz, J. Krell-Roesch, S. Trautwein, A. Scharpf, and A. Woll, "A Tablet-Based App to Support Nursing Home Staff in Delivering an Individualized Cognitive and Physical Exercise Program for Individuals With Dementia: Mixed Methods Usability Study.," JMIR Aging, vol. 6, p. e46480, Aug. 2023, https://doi.org/10.2196/46480.
46. D. N. Kiosses et al., "An Emotion Regulation Tablet App for Middle-Aged and Older Adults at High Suicide Risk: Feasibility, Acceptability, and Two Case Studies," Am J Geriatr Psychiatry, vol. 30, no. 5, pp. 575–584, May 2022, https://doi.org/10.1016/J.JAGP.2021.08.015.
47. L. Appel et al., "mHealth Apps for Dementia, Alzheimer's Disease, and Other Neurocognitive Disorders: A Systematic Search and Environmental Scan", https://doi.org/10.2196/preprints.50186.
48. D. R. Bateman et al., "Categorizing Health Outcomes and Efficacy of mHealth Apps for Persons With Cognitive Impairment: A Systematic Review," J Med Internet Res, vol. 19, no. 8, Aug. 2017, https://doi.org/10.2196/JMIR.7814.
49. H.-L. Kuo, C.-H. Chang, and W.-F. Ma, "A Survey of Mobile Apps for the Care Management of Patients with Dementia.," Healthcare (Basel), vol. 10, no. 7, Jun. 2022, https://doi.org/10.3390/healthcare10071173.
50. T. Engelsma, M. W. M. Jaspers, and L. W. Peute, "Considerate mHealth design for older adults with Alzheimer's disease and related dementias (ADRD): A scoping review on usability barriers and design suggestions," Int J Med Inform, vol. 152, Aug. 2021, https://doi.org/10.1016/j.ijmedinf.2021.104494.
51. S.-Y. Chien, O. Zaslavsky, and C. Berridge, "Technology Usability for People Living With Dementia: Concept Analysis.," JMIR Aging, vol. 7, p. e51987, Jul. 2024, https://doi.org/10.2196/51987.
52. Z. Nimmanterdwong, S. Boonviriya, and P. Tangkijvanich, "Human-Centered Design of Mobile Health Apps for Older Adults: Systematic Review and Narrative Synthesis," JMIR MhealthUhealth, vol. 10, no. 1, Jan. 2022, https://doi.org/10.2196/29512.
53. A.Savoy et al., "Improving Care for People Living With Dementia and Diabetes: Applying the Human-Centered Design Process to Continuous Glucose Monitoring," J Diabetes SciTechnol, vol. 18, no. 1, pp. 201–206, Jan. 2024, https://doi.org/10.1177/19322968221137907.
54. A.M. Petersen et al., "Toward user-centered assistive technologies for aging in place with cognitive impairment: a survey," DisabilRehabil Assist Technol, vol. 19, no. 4, pp. 1561–1567, 2024, https://doi.org/10.1080/17483107.2023.2217848.
55. M. Gomez-Hernandez, X. Ferre, C. Moral, and E. Villalba-Mora, "Design Guidelines of Mobile Apps for Older Adults: Systematic Review and Thematic Analysis," JMIR MhealthUhealth, vol. 11, p. e43186, 2023, https://doi.org/10.2196/43186.
56. C. Berridge et al., "Advance Planning for Technology Use in Dementia Care: Development, Design, and Feasibility of a Novel Self-administered Decision-Making Tool.," JMIR Aging, vol. 5, no. 3, p. e39335, Jul. 2022, https://doi.org/10.2196/39335.

Chapter 11
Smart Healthcare Systems: Enhancing Patient Care Through AI and Big Data

Shelja and Aman Ganesh

Abstract The combination of Artificial Intelligence, Machine Learning, Deep Learning, the Internet of Things, Big Data Analytics, Blockchain, and Augmented Reality/Virtual Reality is transforming healthcare by allowing for real-time monitoring, predictive analytics, improved diagnostics, and secure data management. IoT enables continuous patient monitoring, and AI and ML promote early disease detection and individualized treatments. Blockchain ensures data security, while AR/VR improves medical training and rehabilitation. Despite obstacles such as data privacy concerns, interoperability issues, and high prices, these innovations are transforming patient care, hospital efficiency, and healthcare access. This chapter investigates the effect, benefits, and limitations of these technologies, with case studies that show real-world applications in disease prediction, smart diagnostics, and data security. By giving a detailed overview of new technologies, this book chapter intends to provide useful insights into the future of smart healthcare and its transformational power.

Keywords Healthcare · IoT · Wearable devices · Sensors · Machine learning · Deep learning · Big data

11.1 Introduction

Healthcare plays a crucial role in ensuring overall physical, social, and mental well-being. Advanced technologies have transformed healthcare, enhancing precision, efficiency, and personalized care. Key innovations driving this shift include the Internet of Things (IoT), Artificial Intelligence (AI), Machine Learning (ML), 5G

Shelja
Maharishi Markandeshwar (Deemed to be University), Mullana, Ambala, India

A. Ganesh (✉)
Maharishi Markandeshwar (Deemed to be University), Mullana, Ambala, India
e-mail: aman.ganesh@mmumullana.org

© The Author(s), under exclusive license to Springer Nature Switzerland AG 2026
K. Dalip et al. (eds.), *Healthcare 5.0*, https://doi.org/10.1007/978-3-032-09582-4_11

connectivity, Augmented Reality (AR), Virtual Reality (VR), Big Data Analytics, and Deep Learning.

IoT enables real-time health tracking through interconnected devices, facilitating continuous monitoring and proactive medical interventions. AI and ML process vast amounts of data, uncovering patterns for early diagnosis and personalized treatment. The low-latency, high-speed capabilities of 5G ensure seamless connectivity, crucial for remote patient monitoring and telemedicine. AR and VR enhance medical training and patient therapy, while Big Data Analytics synthesizes complex datasets to improve decision-making and operational efficiency. Deep Learning advances medical imaging and genomics analysis, enabling breakthroughs in disease detection and personalized medicine.

These technologies collectively create an intelligent healthcare ecosystem, optimizing patient care, resource utilization, and medical advancements. The rapid evolution of computer and communication technologies continues to simplify healthcare delivery, making practical solutions more accessible for both researchers and medical practitioners.

11.2 Architecture of Healthcare System: A Brief Insight

Figure 11.1 gives a a brief insight of the layered architecture of the healthcare system which includes basic lyers namely (i) Data collection layer, (ii) network and communication layer, (iii) Data processing layer,. Besides the Architecture also support the management layer and the service layer. These layers form the integral part of the healthcare system This multi-layered foundation for an intelligent healthcare system presented ensures, guaranteeing effective data gathering, processing, administration, and service provision.

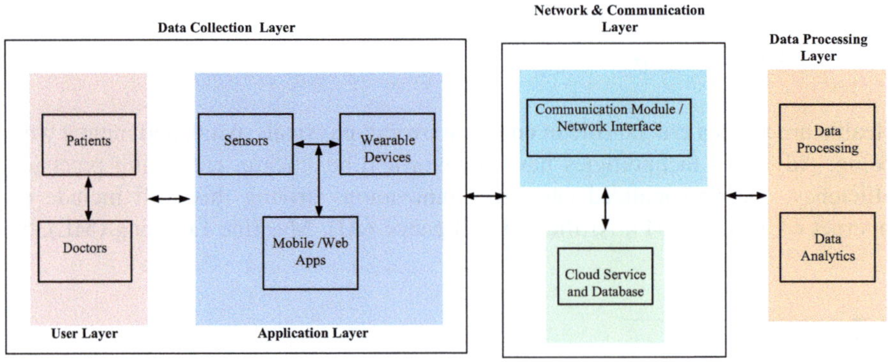

Fig. 11.1 Layered architecture of smart healthcare system

To gather medical data, the data collecting layer combines wearable technology, sensors, and mobile/web apps. The network layer enables secure communication across 4G/5G, WBAN, and Bluetooth. For significant insights, the processing layer uses analytics and data preparation. While the service layerguarantees smooth end-user access, facilitating telemedicine and emergency responses, the management layer organizes healthcare resources. The brief description of each layer is presented below:

(1) *Data Collection Layer*: essential. This layer gathers structured and unstructured medical data from wearable sensors, imaging devices, EHRs, and telehealth platforms. It enables real-time monitoring and data exchange through HL7 and FHIR interoperability standards, ensuring security and compliance with HIPAA regulations. Effective governance at this layer is crucial for predictive analytics and clinical decision support.
(2) *Network and Communication Layer*: Responsible for secure and seamless data exchange, this layer transmits patient records and diagnostic data across healthcare systems. By ensuring real-time connectivity, it enhances telemedicine, collaborative care, and remote monitoring, improving patient coordination and data integrity.
(3) *Data Processing Layer*: This layer aggregates, filters, and analyzes large-scale healthcare data using AI and machine learning to enable predictive analytics and personalized treatments. It enhances decision-making, operational efficiency, and interoperability in data-driven healthcare.
(4) *Management Layer*: Oversees policy implementation, compliance, and resource optimization through hospital information systems (HIS) and management tools. It ensures workflow efficiency, strategic planning, and seamless data exchange, enhancing patient care and operational sustainability.
(5) *Service Layer*: Facilitates end-user interaction for hospitals, ambulances, and emergency services via telemedicine, patient portals, and remote monitoring. It enhances patient engagement, clinical workflows, and proactive healthcare management for better health outcomes.

While the multi-layered architecture provides a structured foundation for smart healthcare, leveraging advanced technologies is essential to maximize its potential. The following section explores the key technological innovations enhancing these layers, their benefits, and the challenges associated with their implementation.

11.3 Technological Advancement in Healthcare System: Review and Challenges

The rapid advancement of technology has transformed healthcare, leading to intelligent systems that utilize blockchain, AI, machine learning, big data, and IoT to enhance patient care, streamline processes, and secure data. Wearable health monitoring devices [3, 4] enhance remote patient monitoring and chronic illness management through real-time tracking, but they face challenges such as high costs, data privacy concerns, and sensor malfunctions. AI-driven diagnostics [5, 6] improve imaging accuracy and reduce diagnostic errors; however, issues like dataset bias and regulatory challenges must be addressed to ensure reliability. Telemedicine and virtual care [7, 8] increase healthcare accessibility, particularly in remote areas, and proved crucial during the COVID-19 pandemic, though they are hindered by internet dependency and the lack of personal interaction in consultations. AI-powered predictive analytics [9, 10] enable early disease detection and personalized treatments, but reliance on historical data and biases in AI models pose challenges. Blockchain technology [11, 12] enhances health data security and transparency, yet scalability and integration with existing systems remain major obstacles. AI-powered personalized healthcare [13, 14] leverages real-time decision-making for tailored treatments, though the complexity of data integration and potential biases must be managed to improve outcomes. 5G technology [15, 16] facilitates real-time remote monitoring and high-quality telemedicine but is limited by infrastructure costs and security risks. IoT in smart hospitals [9, 17] optimizes hospital operations and patient care efficiency, but challenges like data security and outdated system integration persist. Big data analytics [18, 19] drive predictive healthcare models and better decision-making but face hurdles such as privacy concerns and high data processing costs. Augmented reality (AR) [20, 21] improves medical training and surgical precision through real-time interactive models, yet the high cost of AR hardware and user adoption barriers hinder widespread use. Virtual reality (VR) [22, 23] offers immersive training, pain management, and mental health therapy, but its effectiveness is limited by expensive equipment, content creation complexity, and user discomfort. Despite these challenges, these technologies collectively enhance healthcare delivery, improving patient outcomes, efficiency, and accessibility.

Form the above discussion it is can be deduced that IoT enables real-time monitoring and remote care, AI enhances diagnostics and decision-making, big data improves disease prediction and treatment, and blockchain secures medical records. However, challenges remain: interoperability issues hinder data sharing, cybersecurity threats risk patient privacy, AI raises ethical concerns, and high costs limit adoption. Solutions include standardized frameworks for interoperability, stronger cybersecurity, ethical AI guidelines, and cost-effective implementation strategies to enhance accessibility and efficiency in smart healthcare.

11.4 Technological Integration in Smart Healthcare System

To further understand how these technologies interact into healthcare infrastructure, the following sections delve into their role within different layers of a smart healthcare system.

11.4.1 Role of Internet of Things in Different Layers

IoT has transformed rural and urban life through wearable fitness trackers, smart home devices, and smart city programs that enhance traffic flow, energy use, and public safety. In healthcare, H-IoT integrates machine learning (ML) for accurate predictions [24]. Sensors and IoT collect real-time health data via wearables, implanted sensors, and telehealth, enabling remote patient monitoring (RPM) [25]. Smart devices like insulin pumps and inhalers adjust care using real-time data, transmitted via Wi-Fi, Bluetooth, Zigbee, and NFC to cloud networks. Integrated with analytics and Electronic Health Records (EHRs), IoT provides insights for treatment, alerts for timely interventions, and enhances healthcare efficiency.

11.4.2 Role of Artificial Intelligence (AI) in Different Layers

Artificial intelligence AI is transforming healthcare by enhancing diagnosis, treatment, and administrative efficiency [26]. It improves diagnostic accuracy in imaging (MRIs, X-rays) and simplifies data access through machine learning and deep learning [27]. AI automates patient data collection from IoT devices and wearables, utilizing natural language processing (NLP) to extract insights from medical reports [28]. It optimizes network communication by managing capacity, securing data, and enabling low-latency transmission. AI-driven analytics aid disease prediction and personalized treatment. At the management level, AI streamlines administration, integrates with EHRs, and enhances telemedicine through virtual assistants and decision support systems [29], making healthcare more patient-centered and efficient.

11.4.3 Tools Used Under Artificial Intelligence

Numerous AI techniques and technologies are employed in smart healthcare to enhance patient care, diagnosis, treatment, and operational effectiveness. The healthcare system can become more precise, predictive, and individualized thanks to these tools. Key instruments in AI-powered healthcare includes Machine Learning, Deep Learning, and Big data Analytics to name a few.

Machine Learning

Machine learning (ML), a branch of AI, plays a crucial role in disease prediction, aiding early detection and timely treatment [30]. It identifies patterns in vast healthcare data, personalizes therapy, and enhances patient engagement through AI-powered chatbots. ML sorts genomic data, medical imaging, and records to improve diagnosis and predictive analytics, optimizing hospital operations. It also accelerates drug discovery by analyzing genetic and molecular data, supporting personalized medicine. Supervised learning (e.g., random forests, SVMs) aids diagnosis, while unsupervised learning (e.g., PCA, K-means) assists data exploration. Deep learning (CNNs, RNNs, LSTMs) enhances image analysis and predictive modeling, making healthcare more efficient and precise.

Deep Learning

Deep learning enhances smart healthcare by analyzing complex medical data and predicting outcomes [31]. It aids in diagnosing diseases like diabetic retinopathy and skin cancer, often surpassing human expertise. CNNs are widely used for tumor detection, while RNNs and LSTMs predict disease progression. GANs improve cancer detection accuracy and generate synthetic medical data for robust training [32, 33]. Transfer learning further enhances medical AI by leveraging pre-trained models. These deep learning techniques advance diagnostics, treatment planning, and personalized medicine, optimizing healthcare efficiency.

Big Data Analytics

There are many factors that are considered in performing big data analysis, some of them are: the patient's medical history, genetic information, eating habits and fitness regimen. The data that is analyzed includes several key decision-making processes. Some of the challenges with the data used include data quality, data validation, data knowledge, domain expertise, and data integration challenges with various end points. While performing data analysis, the HealthCare Sectors must take security and data governance (HIPPA regulations etc.) into consideration. There are many factors that are considered in performing big data analysis, some of them are: the patient's medical history, genetic information, eating habits and fitness regimen. The data that is analyzed includes several key decision-making processes. Some of the challenges with the data used include data quality, data validation, data knowledge, domain expertise, and data integration challenges with various end points. While performing data analysis, the HealthCare Sectors must take security and data governance (HIPPA regulations etc.) into consideration.

Big data analytics in smart healthcare involves collecting and analyzing diverse datasets from wearables, medical imaging, genomics, and EHRs to enhance care quality [34]. Based on sample size (n) and parameters (p), healthcare big data falls into three categories: small-large, large-small, and large-large. Factors like medical history, genetics, and lifestyle influence analysis, with concerns about data quality, validation, integration, and security (HIPAA laws). AI and ML refine raw data for personalized care, disease prevention, detection and treatment. Key algorithms include MapReduce for large-scale data processing, Hadoop for scalable analysis, and Spark for real-time applications. These enable predictive modeling and evidence-based decision-making in healthcare.

11.5 Case Studies

To better understand the application of above mentioned technologies in real world, their validation in terms of theoretical concepts, showcasing practical applications of smart healthcare technologies while demonstrating measureable impact case studies are included in this section. By the means of case studies credibility, applicability, and future research directions in smart healthcare innovations can be understood.

11.5.1 Case Study: Machine Learning for Early Predication of Diabetes

Victor Chang et al. [40] evaluated multiple ML models for early diabetes prediction using key performance indicators. They used a dataset with patient health records, including BMI, blood glucose levels, blood pressure, and cholesterol for the predicition.

The study compared various machine learning algorithms, revealing Random Forest (RF) and K-nearest Neighbor (KNN as the most accurate and robust, making them suitable for clinical applications in comparison to Naive Bayesian (NB), Logistic Regression (LR) and Decision Trees (DT). It also highlighted the importance selection of the feature in improving model efficiency. The performance metrics for the above mentioned machine models is recalled and tabulated in Table 11.2 and Fig. 11.2.

The study emphasized the potential of machine learning (ML) in diabetes prediction, highlighting its potential to improve early intervention strategies and reduce long-term complications and healthcare costs. However, the authors also highlighted the challenges like data imbalance and interpretability which required further research.

Table 11.2 Performance metrics for ML model

Classifier	Accuracy (%)	Precision (%)	Senstivity (%)	Specificity (%)	F1-score (%)	Hamming loss (%)
DT	81.12	83.01	77.93	84.04	81.11	18.97
RF	82.25	83.45	80.47	84.06	82.25	17.73
KNN	80.54	81.05	79.51	81.58	80.53	19.43
LR	72.63	72.07	73.96	71.33	72.63	27.35
NB	70.57	72.05	67.05	74.03	70.53	29.48

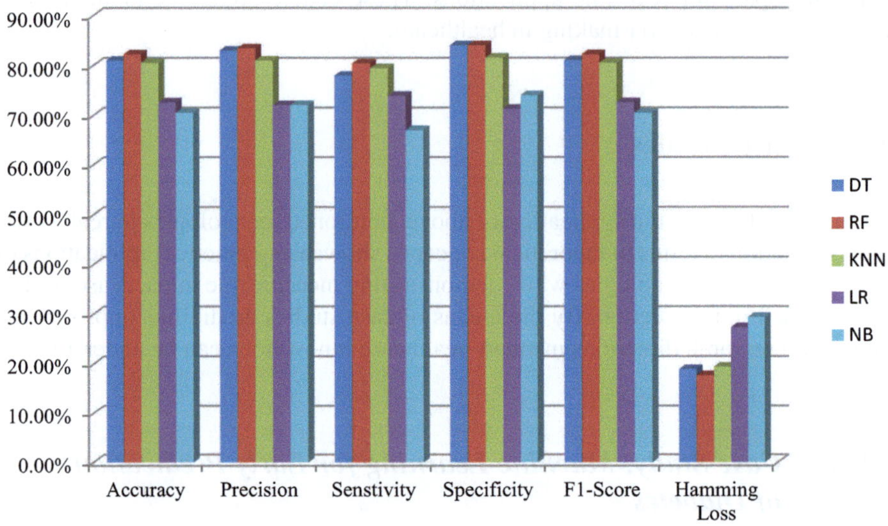

Fig. 11.2 Comparison of ML model performance

11.5.2 Case Study: Deep Learning for Lung Cancer Detection Using 3D Thoracic MRI Scan

Li et al. [41] proposed a deep learning-based method for lung nodule detection in 3D thoracic MRI scans, addressing early lung cancer diagnosis challenges and optimizing R-CNN detection through spatial three-channel input construction.

The study uses 142 T2-weighted MR scans from the First Affiliated Hospital of Guangzhou Medical University consisting of 142 slices with 7.7 mm spatial resolution. The lung nodules can be divided into isolated, juxta vascular, and juxta pleural nodules. The same are shown in Fig. 11.3.

A Faster R-CNN technique was employed to improve lung nodule identification. The R-CNN module consisted of region Proposal Network (RPN) employed for generating proposed region for the selected lung nodule image. This was followed by Fast R-CNN for classification through feature extraction using ResNet50 and

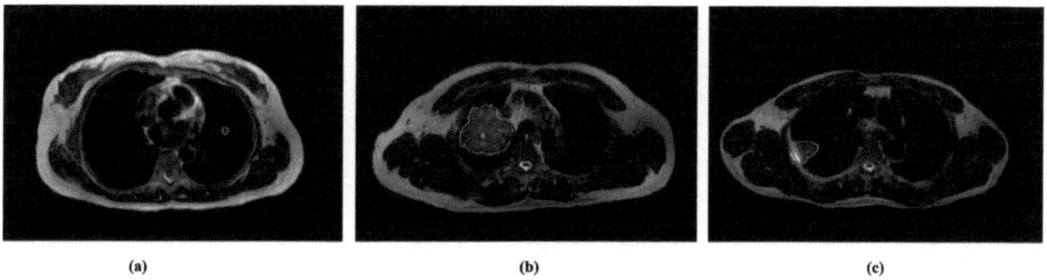

Fig. 11.3 Example of **a** an isolated lung nodule **b**) juxta vascular nodule **c** juxta pleural nodule [41]

ResNet100. Both the module shared the convolution layers The effectiveness of Fast R-CNN was carried out against local ternary patterns (LTP), Histogram of Gradient (HOG), handcrafted CT images (H-CT), CAD based CT images (CAD-CT), shallow CNN based CT images (CNN-CT) and X-Ray based CNN (X-CNN). A three-channel spatial input distinguishes lung nodules from air artifacts, and an FP reduction scheme ensures detection reliability. The method achieved 85.2% sensitivity with 3.47 false positives per scan, significantly enhancing lung nodule identification. A tabular view of the detected nodule for the selected dataset along with pictorial comparison is shown below in Table 11.3 and Fig. 11.4 respectively where FPs/scan represent the false positive/scan.

The study confirmed the superiority of the proposed deep learning approach in MRI-based detection of lung cancer, despite challenges in detecting small and low-contrast nodules, highlighting the potential for improved, non-invasive diagnostic tools in medical imaging.

Table 11.3 Comparison of nodule detection

	Sensitivity = 80%, FPs/scan	Sensitivity = 85%, FPs/scan
LTP	14.7	17.4
HOG	16.3	19.8
H-CT	17.8	21.7
CAD-CT	18.7	20.7
CNN-CT	22.4	25.2
X-CNN	14.8	19.9
Fast R-CNN	**2.38**	**3.47**

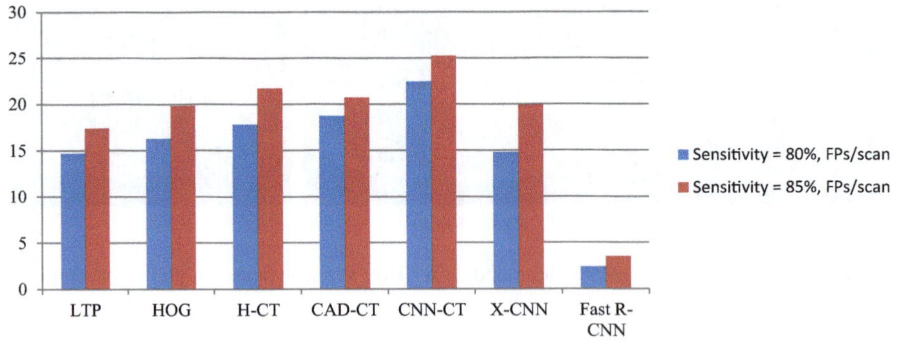

Fig. 11.4 Comparison of CNN model performance

11.5.3 Case Study: Big Data for Forecasting Disease Outbreak

Sujata Dash et al. (2021) used the Big-Data Driven Intelligent Forecasting Model (BIFM) to predict COVID-19 outbreaks using advanced time-series techniques. They used the ARIMA model to analyzed future COVID-19 spread in six countries and 21 Indian states. The primary objective was to develop accurate short-term predictions that can assist in healthcare planning and resource allocation.

The ARIMA model, was employed and this used historical COVID-19 data made available from from https://www.covid19india.org/ and Johns Hopkins University Coronavirus Data Stream for six countries viz. US, Brazil, India, France, United Kingdom and Russia and to predict state-wise disease status of Indai. The datasets were imported using CSV format, combining WHO and CDC case data.

ARIMA is forecasting model that uses past observations to predict future values, defined by parameters like p, d, and q, which include past forecast errors where p is number of past observations (lags) used to predict the future, d is number of times past data points are differenced to make the series stationary and q is number of past forecast errors included in the model. The ARIMA model was employed for accurately predicting the daily –confirmed cases for the next 90 days. The best fitted ARIMA model for selected countries is tabulated in Table 11.4.

Referring to Table 11.4, The ARIMA model effectively predicts COVID-19 trends across regions, with optimized ARIMA order values and low AIC, BIC and RMSE values, confirming its reliability and statistical significance, making it suitable for public health decision-making. The same is shown in Fig. 11.5. The low p-values further reinforce the statistical significance of the predictions, making them suitable for public health decision-making.

Similarly for Indian States the comparative score is shown in Fig. 11.6.

Table 11.4 ARIMA model for the top six worst affected countries, with confirmed cases

Sl. no	Country	ARIMA order	AIC	BIC	RMSE predicted
1	US	(0,1,1)	9751.353	9759.489	17,491.107
2	Brazil	(2,1,1)	9403.853	9428.302	18,813.727
3	India	(2,2,1)	8464.357	8488.811	5260.501
4	France	(2,1,1)	9156.730	9181.187	9870.122
5	Russia	(0,2,2)	6747.133	6759.266	565.881
6	United Kingdom	(2,1,1)	7978.710	8003.139	3603.24

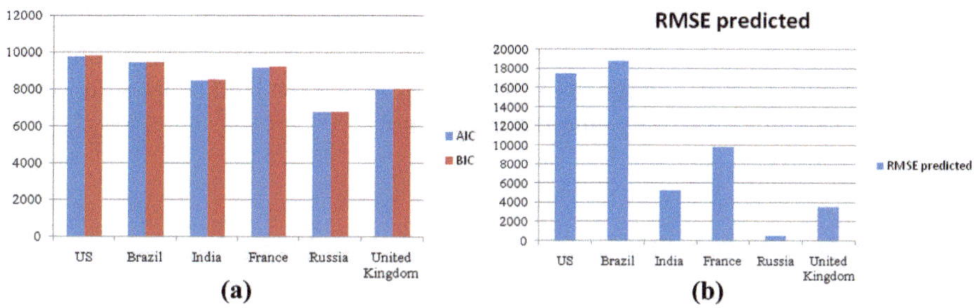

Fig. 11.5 ARIMA model performance for affected countries

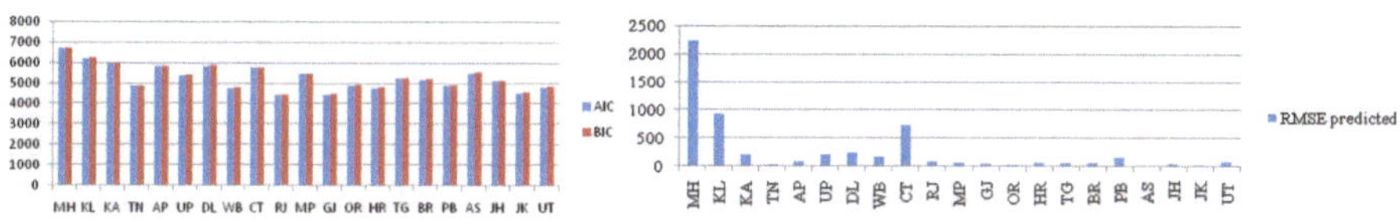

Fig. 11.6 ARIMA model performance for affected Indian states

The ARIMA model accurately predicted daily-confirmed cases for the next 90 days, achieving an 85% Mean Absolute Percentage Error. The study also evaluated the impact of lockdown measures on disease spread in India, identifying trends and potential healthcare system constraints.

The significance of data-driven forecasting in managing pandemics and future healthcare crises is highlighted by the study, which suggests integrating big data analytics with machine learning models could enhance COVID-19 prediction accuracy and decision-making for public health interventions.

11.6 Challenges in Smart Healthcare

Despite its potential, smart healthcare faces key challenges. Data security and privacy remain critical, requiring protection from cyber threats and compliance with HIPAA and GDPR. Interoperability issues hinder seamless communication among systems, while high costs limit accessibility, particularly in underserved regions. Managing vast healthcare data demands robust infrastructure and advanced analytics to derive meaningful insights.

Additionally, user acceptance and training are essential, as healthcare professionals and patients may resist new technologies due to unfamiliarity. Regulatory and ethical concerns, such as compliance issues and AI-driven biases, must be addressed to ensure trust and equity. Many hospitals still rely on legacy systems, making integration difficult. Lastly, continuous validation of AI-driven healthcare solutions is crucial for accuracy, reliability, and patient safety.

11.7 Future Scope: Emerging Computing Technologies

Advancements in AI, predictive analytics, and telemedicine will enable real-time health monitoring and remote care. Blockchain technology will enhance data security, administrative processes, and workflow efficiency. Automation and robotics will improve surgical precision, while 3D printing will revolutionize organ transplants and medical equipment manufacturing.

Nanotechnology holds promise for targeted drug delivery, improving treatment outcomes. Augmented and virtual reality will enhance medical education, diagnostics, and rehabilitation, while CRISPR and genomic medicine will enable gene-editing therapies for genetic disorders. Integrated health systems will unify wearable, hospital, and lab data for better clinical decision-making. Green healthcare solutions will drive sustainability through eco-friendly waste management and energy-efficient hospital designs.

The future of healthcare is vast and transformative, with emerging technologies making it more personalized, efficient, and accessible than ever before.

11.8 Conclusion

The integration of emerging technologies into healthcare has led to a smarter, more efficient, and patient-centric system. These technologies include IoT, AI, Machine Learning, Deep Learning, Big Data Analytics, Blockchain, Augmented Reality (AR), and Virtual Reality (VR). They enhance real-time monitoring, predictive analytics, and decision-making. IoT-enabled wearable devices and wireless sensors enable continuous patient monitoring, while AI-driven analytics provide early disease detection, personalized treatment, and automated decision-making. Big Data Analytics manages vast healthcare datasets, optimizes resource allocation, and improves diagnostic accuracy. However, challenges like interoperability, data security, ethical dilemmas, and integration with legacy systems remain. Addressing these challenges through standardized healthcare frameworks, enhanced cybersecurity measures, and cost-effective implementation strategies is crucial for unlocking the full potential of smart healthcare.

References:

1. G. Salloum and J. Tekli, "Automated and personalized nutrition health assessment, recommendation, and progress evaluation using fuzzy reasoning," *Int. J. Hum.-Comput. Stud.*, vol. 151, Jul. 2021, Art. no. 102610.
2. T. Arora and I. Grey, "Health behavior changes during COVID-19 and the potential consequences: A mini-review," *J. Health Psychol.*, vol. 25, no. 9, pp. 1155–1163, Aug. 2020.
3. M. S. Patel et al., "Wearable technology for health monitoring: A review," *Health Informatics J.*, vol. 23, no. 4, pp. 259–270, 2017.
4. A. Esteva et al., "Dermatologist-level classification of skin cancer with deep neural networks," *Nature*, vol. 542, no. 7639, pp. 115–118, 2017.
5. P. Rajpurkar et al., "CheXNet: Radiologist-level pneumonia detection on chest X-rays with deep learning," *arXiv preprint*, arXiv:1711.05225, 2017.
6. R. L. Bashshur, G. Shannon, N. Bashshur, and P. M. Yellowlees, "Telemedicine and e-health," *Telemedicine and e-Health*, vol. 22, no. 10, pp. 789–794, 2016.
7. T. Greenhalgh et al., "Virtual consultations: Advantages and limitations," *BMJ Innovations*, 2019.
8. E. Choi et al., "Doctor AI: Predicting clinical events via recurrent neural networks," *IEEE Trans. Biomed. Eng.*, 2020.
9. M. I. Razzak et al., "Big data analytics for smart healthcare systems: A survey," *IEEE Access*, vol. 8, pp. 95714–95733, 2020.
10. C. C. Agbo et al., "Blockchain technology in healthcare: A comprehensive review and directions for future research," *Appl. Sci.*, vol. 11, no. 10, 2021.
11. X. Liao et al., "Blockchain-based secure sharing of personal health records," *Future Gener. Comput. Syst.*, vol. 124, pp. 75–86, 2021.
12. E. J. Topol, "AI in healthcare: The next frontier in personalized medicine," *J. Pers. Med.*, vol. 12, no. 1, 2022.
13. A. Shah et al., "Artificial intelligence in personalized healthcare: Advancements and challenges," *Comput. Biol. Med.*, vol. 146, 2022.
14. Zhang et al., "5G in healthcare: A survey and future directions," *IEEE Access*, vol. 11, pp. 11245–11259, 2023.
15. X. Li et al., "The role of 5G in the future of telemedicine," *IEEE J. Sel. Areas Commun.*, vol. 41, no. 2, pp. 125–138, 2023.
16. W. Chien et al., "Smart healthcare: IoT-based solutions for healthcare applications," *Sensors*, vol. 24, no. 3, 2024.
17. S. Kumar et al., "IoT and big data analytics for smart hospitals: A review," *Healthc. Technol. Lett.*, vol. 11, no. 1, pp. 23–34, 2024.
18. 19. M. Asiri et al., "A comprehensive survey on big data analytics for healthcare applications," *IEEE Access*, vol. 9, pp. 73574–73588, 2021.
19. 20. Z. Zheng et al., "Augmented reality in healthcare: A review of applications and future directions," *IEEE Access*, vol. 9, pp. 110456–110472, 2021.

20. 21. S. Monfared et al., "Application of augmented reality for medical training and surgical simulation," *J. Med. Syst.*, vol. 46, no. 2, pp. 45–58, 2022.
21. 22. J. Munafo et al., "Virtual reality in healthcare: Applications, benefits, and challenges," *IEEE Trans. Med. Robot. Bionics.*, vol. 4, no. 1, pp. 75–89, 2022.
22. 23. M. Shah et al., "The potential of virtual reality in rehabilitation: A systematic review of applications in therapeutic contexts," *J. Rehabil. Res. Dev.*, vol. 59, pp. 101–115, 2022.
23. 24. R. Beecham et al., "A review of VR applications in healthcare," *J. Healthc. Eng.*, vol. 18, no. 1, pp. 87–102, 2023.
24. 25. H. K. Bharadwaj et al., "A Review on the Role of Machine Learning in Enabling IoT Based Healthcare Applications," *IEEE Access*, vol. 9, pp. 38859–38890, 2021, https://doi.org/10.1109/ACCESS.2021.3059858.
25. M. M. H. Mia, N. Mahfuz, M. R. Habib, and R. Hossain, "An Internet of Things Application on Continuous Remote Patient Monitoring and Diagnosis," *2021 4th Int. Conf. Bio-Eng. Smart Technol. (BioSMART)*, Paris / Créteil, France, 2021, pp. 1–6, https://doi.org/10.1109/BioSMART54244.2021.9677715.
26. S. Nagar et al., "Review and Explore the Transformative Impact of Artificial Intelligence (AI) in Smart Healthcare Systems," *2024 Int. Conf. Adv. Comput. Res. Sci. Eng. Technol. (ACROSET)*, Indore, India, 2024, pp. 1–5, https://doi.org/10.1109/ACROSET62108.2024.10743527.
27. T. Wu and T. Zhang, "Research on Intelligent Extraction Technology of Unstructured Test Data," *2024 6th Int. Conf. Internet Things, Autom. Artif. Intell. (IoTAAI)*, pp. 575–579, 2024.
28. 29. B. Zhou, G. Yang, Z. Shi, and S. Ma, "Natural Language Processing for Smart Healthcare," *IEEE Rev. Biomed. Eng.*, vol. 17, pp. 4–18, 2024, https://doi.org/10.1109/RBME.2022.3210270.
29. J. Jayapradha et al., "A Telehealth System Driven by Artificial Intelligence for Effective Patient Consultation and Diagnosis in Hospitals," *2023 Int. Conf. Syst. Comput. Autom. Netw. (ICSCAN)*, Puducherry, India, 2023, pp. 1–6, https://doi.org/10.1109/ICSCAN58655.2023.10395119.
30. A. Bisht, D. Dhiman, and N. Yamsani, "Making Health Care Smarter: The Role of Machine Learning Algorithms," *2023 6th Int. Conf. Contemp. Comput. Informat. (IC3I)*, Gautam Buddha Nagar, India, 2023, pp. 919–924, https://doi.org/10.1109/IC3I59117.2023.10397900.
31. P. Juyal, "Enhancing Predictive Analytics in Healthcare Leveraging Deep Learning for Early Diagnosis and Treatment Optimization," *2024 5th Int. Conf. Smart Electron. Commun. (ICOSEC)*, Trichy, India, 2024, pp. 1988–1993, https://doi.org/10.1109/ICOSEC61587.2024.10722504.
32. K. S. Gajjala and B. Chakraborty, "Human Activity Recognition based on LSTM Neural Network Optimized by PSO Algorithm," *2021 IEEE 4th International Conference on Knowledge Innovation and Invention (ICKII)*, Taichung, Taiwan, 2021, pp. 128–133, https://doi.org/10.1109/ICKII51822.2021.9574788.
33. S. Gupta, V. Pushparajesh and J. M. Sahoo, "Evaluating the Performance of Generative Adversarial Networks for Cancer Detection in Smart Healthcare Systems," *2024 15th International Conference on Computing Communication and Networking Technologies (ICCCNT)*, Kamand, India, 2024, pp. 1–6, https://doi.org/10.1109/ICCCNT61001.2024.10724224.
34. 35. S. Nazir *et al.*, "A Comprehensive Analysis of Healthcare Big Data Management, Analytics and Scientific Programming," in *IEEE Access*, vol. 8, pp. 95714-95733, 2020, https://doi.org/10.1109/ACCESS.2020.2995572.
35. 36. A. Alnuaimi, D. Hawashin, R. Jayaraman, K. Salah and M. Omar, "Trustworthy Healthcare Professional Credential Verification Using Blockchain Technology," in *IEEE Access*, vol. 11, pp. 109669-109688, 2023, https://doi.org/10.1109/ACCESS.2023.3322359.
36. A. Y. Bayahya, W. Alhalabi, and S. H. AlAmri, "Smart health system to detect dementia disorders using virtual reality," *Healthcare*, vol. 9, no. 7, p. 810, Jun. 2021.
37. A. Williams et al., "The use of algorithms in AR/VR healthcare applications: A comprehensive review," *Int. J. Med. Robot. Comput. Assist. Surg.*, 2023.
38. R. Smith et al., "Advancements in VR rendering algorithms for healthcare applications," *J. Med. Imaging VR*, 2024.

39. T. Nguyen et al., "The role of VR and AR in surgical precision and medical training," *J. Surg. Robot. Comput. Assist. Surg.*, 2023.
40. Victor Chang, Meghana Ashok Ganatra, Karl Hall, Lewis Golightly, Qianwen Ariel Xu, "An assessment of machine learning models and algorithms for early prediction and diagnosis of diabetes using health indicators" *Healthcare Analytics*, vol. 2, no. 8, 2022,
41. 42. Y. Li, L. Zhang, H. Chen and N. Yang, "Lung Nodule Detection With Deep Learning in 3D Thoracic MR Images," in *IEEE Access*, vol. 7, pp. 37822-37832, 2019, https://doi.org/10.1109/ACCESS.2019.2905574.
42. 43. S. Dash, C. Chakraborty, S. K. Giri, S. K. Pani and J. Frnda, "BIFM: Big-Data Driven Intelligent Forecasting Model for COVID-19," in *IEEE Access*, vol. 9, pp. 97505-97517, 2021, https://doi.org/10.1109/ACCESS.2021.3094658.

Chapter 12
Hair Disease Detection System Using Machine Learning

Srishti, Dalip, and Navjeet Kaur

Abstract One of the most common diseases which acts as normal but poses a highest crucial role in dermatology which requires fast diagnosis and an efficient therapy. In the traditional times, the dermatologist's approach is based on the subjective diagnosis only which only include the visual inspections which can be more time consuming and prone to errors. As with the enhancement in the technologies, and the world's cycle goes more with Artificial intelligence and Machine Learning approaches. So, in this paper we are using the Machine Learning approaches for the automation of detection procedure using different ML algorithms and Deep learning approaches to get the desired results which provide accurate and trustworthy results in an early stages. By using these methods, the machines aims' to provide the better or accurate diagnosis, accurate results about the hair disease detection which is better than visual inspection and hence results in less time consumption for the dermatologists. The proposed model includes the multi stage techniques include data preprocessing, getting the best feature extraction using different ML techniques and then classify the data to detect different hair disorders. Deep learning methods such as CNN (Convolutional Neural Networks) is also used for detection based on the visuals datasets to achieve the accuracy. The system should be trained and tested using wide range of images. By assisting dermatologists in correctly diagnosing hair issues, this approach hopes to enhance patient care and treatment results.

Keywords Diagnosis · Image processing · Convolutional neural networks (CNNs) · Hair diseases · Machine learning (ML)

Srishti
Guru Nanak Institute of Technology, MULLANA, Ambala, Haryana, India

Dalip · N. Kaur (✉)
M.M. Institute of Computer Technology and Business Management, Maharishi Markandeshwar (Deemed to Be University), MULLANA, Ambala, Haryana, India
e-mail: navjeet.kaur@mmumullana.org

© The Author(s), under exclusive license to Springer Nature Switzerland AG 2026

K. Dalip et al. (eds.), *Healthcare 5.0*, https://doi.org/10.1007/978-3-032-09582-4_12

12.1 Introduction

For Humans, hair plays a vital role in the life, both biologically and culturally, but it will be depending on everyone personal prospective and cultural contexts. Hair is an important part of human body as hair protects the scalp from environmental elements, such as the sun's UV rays, dust, and harsh weather conditions. Same as Eyelashes protect the eyes from debris, and eyebrows help shield the eyes from sweat. Hairs is a significant part of one's identity. Healthy and well- maintained hairs symbolize one's good health and attractiveness. On the other-hand one can feel self-esteem, more confident when their hairs meet their personal standards.

The condition which can affect a person's physical appearance, mental health, and general quality of life create a huge impact on the person's life which can be caused by the simple hair loss or we can also termed it as a hair disease. The very simple issues like alopecia to more complex diseases like folliculitis and trichotillomania [1]. These conditions, which can range from simple issues like alopecia to more complex disorders like folliculitis and trichotillomania, can have a significant impact on prompt and accurate diagnosis, which is essential for the best possible care and management of hair illnesses. However, traditional diagnostic techniques typically rely on dermatologists' subjective visual assessments, which cause variation in diagnosis and treatment delays. The development of Machine Learning (ML) techniques in recent years has created new avenues for automating and enhancing the diagnosis of hair disorders and almost it works more efficiently in every field not only in medical but also in the field of market analysis, automation, neural networks and data mining etc. The Machine Learning is generally provide us the solutions of the problems that are prediction based and the general data. In our research, the Machine Learning basically provides the increased accuracy rates of the developed models so that the models works best for the early detection of diseases and get better outputs. It mainly works with the high amount of complex data and categorial type to data so that they can get better results. The skin and scalp images that are used through computer vision techniques to design machine learning model using CNN model to analyze the scalp and hairs. The developed model then can analyze images to detect various health issues like seborrheic dermatitis, psoriasis, tinea capitis, and alopecia. Huge datasets containing images of both healthy and unhealthy scalp conditions can be used to train or test machine learning algorithms, enabling them to identify unique features and patterns interrelated with each condition to specify different types of hair loss, scalp inflammation, fungal infections, or other hair diseases based on visual patterns, making diagnosis faster and more consistent.

Clinical professionals can then take advantage of these algorithms' ability to automatically categorize and identify hair problems based on input data, providing valuable information and enabling timely interventions. There is great promise for improving patient outcomes and advancing the field of dermatology with the integration of Machine Learning into hair disease detection systems. By detecting the early diagnosis of the hair diseases, early hair disease treatments and their improved detection accuracy rates would being helped in the better treatments [2]. In addition to these technologies, it would be greatly helped to achieve the target of improved skills of the healthcare professionals in their vital decision-support tools and increase the efficiently in their personalized treatments. In the new digital era, the use of ML based models will provide the easy or efficient access to the hair disease detections and will provide the unique ways to get the early predictions about the diseases and get the more accurate or better results by continuously monitoring and check the progress through tracking. Furthermore, these technologies can use the data-driven studies for data preprocessing and then generate accumulative results based on better understanding and tailored results on different therapies. Overall, the combination of Machine Learning and dermatology has n number of benefits.

12.2 Literature Review

In order to diagnose a range of hair issues, numerous investigations will be conducted on image classification approaches employing various visuals based on distinct color factors or varied viewpoints. Smith and colleagues (2018) study has been achieved good classification accuracy rates by using CNNs to classify images of several scalp-based conditions, including psoriasis, alopecia, and seborrheic dermatologist [1] to achieve the accuracy. Similarly, on the other hand a deep learning-based system has also been automated for scalp condition diagnosis was developed by Jones et al. (2020) and it also helps to get early diagnosis demonstrated in differentiating between different types of hair issues. In order to extract better results and classify different results they classify pictures using Machine Learning to extract the best possible features using feature extraction method. A wavelet transform and Grey-Level Co Occurrence Matrix (GLCM) based feature extraction strategy was proposed in previous researches to identify hair diseases from scalp pictures, and it performed similarly to earlier approaches to get the accurate predictions.

It has been shown in the research by Mrinmoy Roy and Anica Tasnim Protity that there will be total of 1,000,00 healthy hairs which are present on the human scalp and it is absolutely normal to loose 50–100 strands of hairs on the daily basis that will not be considered as hair fall or hair loss [3]. Hair loss is not a modern problem. Common causes of hair loss include psoriasis, folliculitis, and alopecia. Alopecia differs from regular hair loss in that the latter causes coin-sized bald patches all over the scalp area. There are various types of alopecia or patchy hair loss. The most common type of alopecia, also known as male-pattern baldness (MPB), is androgenetic alopecia, in which the hairline begins to recede in a pattern that primarily affects the frontal and temple areas. 70% of men and 40% of women experience this type of hair loss and thinning problems.

Hair thickens by 50% when applied topically with minoxidil solution [4]. Alopecia areata (AA), on the other hand, is an autoimmune condition that affects people of any age or gender. Although AA mostly affects the scalp, it can also spread to the beard, eyebrows, and eyelashes. In this instance, hair follicles are not recognized as "self" by the body's immune cells. Rather, they view these follicles as "foreign," which ultimately leads the immune cells to target and eliminate the hair follicles. The study from [5] reported that, in the US alone, 700,000 individuals suffer from AA. This disease, if diagnosed early, might resolve spontaneously. In severe cases, topical corticosteroid or immune therapy is used. In order to solve the difficulties, they created a model that has a low false-positive and false-negative rate and can accurately diagnose psoriasis, folliculitis, and alopecia. Even though gathering photographs of the conditions from the internet can be difficult because the images vary in color, shape, and resolution, they used a variety of pre-processing techniques, including denoising, scaling, and augmentation, to produce a dataset that could aid in future study on scalp diseases.

The best hyper parameters chosen from the grid search were used to train the CNN model. The dataset was split into 70%-30% train-test-split, with 45 randomly chosen images used for testing and 105 randomly chosen images used for training. Following the pre-processing procedures, they trained the CNN model on the training dataset and utilized the model to assess the test dataset. On unseen data, their system achieved 96.2% training accuracy and 91.1% validation accuracy. From epoch 1 to epoch 50, training losses decreased from 1.1685 to 0.1017, and validation losses decreased from 1.1260 to 0.3438. At the same time, training accuracy and validation accuracy increased to 96.2% and 91.1%, respectively.

Another study of Sheshang Degadwala and Dhairya Vyas they are using the three widely utilized Convolutional Neural Network (CNN) architectures—AlexNet, VGG16, and ResNet50—this study explores the use of transfer learning to leverage a multiclass categorization of hair diseases. In the early stage of their research their first step is to get the large dataset of high quality pictures (based on resolution, brightness, contrast etc.) based on different hair scalp or hair conditions such as androgenetic alopecia, tinea capitis, and alopecia areata. Then a comprehensive study has been carried out to develop the models and made it possible to increase the accuracy rate from the dataset based on different kinds and their conditions.

The hair scalp dataset of different diseases is used to adjust these neural network designs using a transfer learning technique. In order to increase the accuracy of the model's performance, a hyperparameter tuning approach is also used to fine-tune different parameters, including learning rates, batch sizes, and optimization techniques. The research would result in all three architectures—including ResNet50—classify multiclass hair disorders with an accuracy rate of 99% [6].

The use of ensemble learning techniques—which is one the reliable prediction technique and can be used to make accurate predictions using multiple models. This technique can be used for the detection of hair disorders. In order to detect one of the hair problem named alopecia areata, Wang et al. built an ensemble model that performed better than individual classifiers and includes Support Vector Machines (SVMs), random forests, and K-Nearest Neighbors (KNN). In a same manner, [7] proposed an ensemble-based method that combined CNNs and decision trees for scalp condition classification, to clearly show the robustness and cross-dataset generalization of the dataset. In the developing and testing of ML-based hair disorder detection systems it requires high-quality datasets and standard procedures. In the training and testing stage, the researchers used quite a lot of publically available datasets, such as the International Skin Imaging Collaboration (ISIC) and the Hair Disease Image Dataset (HDID) to get the desired output. Moreover, well-known benchmarking methods like holdout validation and cross-validation, have also been used to calculate suggested models' performance and ensure their consistency across studies. Regardless of advancements, limitations and impediments to ML-based hair illness detection still exist. These include issues including class imbalance, dataset bias, and generalization to other patient populations. Moreover, the accountability of machine learning models and their consolidation into clinical practice becomes the primary areas of research and development.

12.3 Existing System

Data processing refers to the approaches used to filter out the data so that any valuable information would be filtered to get the better or accurate results. Image preprocessing is also one of its crucial step. As we are using different ML techniques and CNN techniques which require a lot of visual datasets to get the accuracy based on the visuals. It requires different approaches to get the improved quality images. These methods include equalization, noise reduction, reflection removal, and intensity modifications. The target of image processing is to improve human perception and input for automated systems. These techniques reduce distortions and provide attention to particular visual characteristics in order to prepare photographs. They do this by using low-pass filters to reduce high-frequency noise while maintaining important image properties.

Convolutional neural networks (CNNs): Convolutional neural networks, or CNNs, have emerged as a crucial component of deep learning for applications such as computer vision and speech recognition. Because CNNs use parameter sharing, they are easier to use and comprehend in hierarchical feature extraction than fully linked networks. By lowering saturation issues, ReLU and other nonlinear activation functions increase accuracy at a lower cost of problem-solving. By using hyperparameters like filter size and stride, pooling layers can construct efficient feature extraction and minimize dimensionality while maintaining crucial information. Fully connected layers facilitate thinking by connecting neurons and offering complex transformations and higher-level abstraction. CNNs also provide a powerful foundation for image processing, allowing for reliable and effective solutions in a variety of applications to achieve precise outcomes.

Seo et al. One of the most common hair disease called alopecia areata, their number of patients has recently been enlarged due to the unhealthy eating habits and the various environmental factors. In this study, we provide an algorithm for minimizing hair loss and scalp self-diagnosis that extracts HLF (hair loss feature) from scalp photos via a microscope mounted on a smart device. We extract the HLF by combining a scalp image recorded under a microscope with grid line selection and eigenvalue calculation. First, we utilize image processing to improve the microscope input contrast and eliminate light reflection in scalp photographs. Second, HLF is obtained using each distinct approach to determine the degree of hair loss based on the preprocessed scalp image.

Ibrahim et al Scalp issues can result from a variety of variables, including heredity, stress, abuse, and hair products. The conventional approach of treating the scalp and hair is costly and difficult to diagnose. Furthermore, payers must increasingly evaluate the efficacy of novel therapy options in the management of a specific scalp condition. Because they are often costly and inconvenient, there is an increasing demand for a less expensive and simpler way to monitor scalp disorders. Thus, this study includes an examination into the pre-trained classification of scalp illnesses using image processing techniques.

12.4 Proposed System

A proposed approach for diagnosing hair illnesses may combine technology, medical skill, and data analysis. The methods for data analysis are given below:

Input Data Set: Our dataset is designed to help a machine learning system detect hair diseases accurately. It includes a variety of high-quality images showing different hair conditions, collected from trusted medical sources like dermatology databases and hospitals. These images cover a wide range of hair disorders, including alopecia, fungal infections, hereditary hair loss, scalp conditions, and more. The data has been collected for the data pre-processing through various patients disorders and different pictures has been taken for further processing. The different types of pictures shows different types of diseases and its causes and, in some cases, information like the patient's age, gender, medical history, and environmental factors or concerns that could affect their hair health and quality. The techniques which we are using provide a detailed review of the real world situations and the factors dependencies on different diseases [8]. The images we are using are collected from different sources and the data from different studies of different researchers. For the training, we prepare our images from different techniques. We resize or adjust them to keep their dimensions consistent, adjust pixel values for better learning, and use data augmentation methods or techniques like flipping, rotating, and adjusting contrast to make the model more adaptable and reliable. CNN based advanced image processing technique is also used to focus on the hair and scalp areas while reducing background noise using different techniques. During the whole process, we are assuring the ethical standards (Code of Medical Ethics Regulations, 2002) and preparing the whole documentation. By carefully collecting, labelling, and refining the data, we aim to createa strong machine-learning based model that can accurately identify and analyze the hair-related conditions and problems.

Pre-processing: Once the data has been collected our next phase in the building of machine learning model is to do the pre-processing of data. We has been collected the data from different sources. Now, we use different algorithms for the data pre-processing. The pictures we are collecting for our research are of high quality which are then categorized on the basis of their normalization, scaling, and augmentation during pre-processing for a machine learning-based model for hair disease identification system to ensure consistency and flexibility. The data will be handled with the honest considerations that mist be ensured with full secrecy and patient authority [9]. The research findings are clear, flexible and replicable thanks to the detailed records that come with the dataset. To achieve properties like consistency, robustness, and accuracy, pre-processing is one of the essential step in preparing the dataset for a machine learning-based model of hair disease detection system. To achieve the consistent brightness and contrast features, high quality resolutions pictures has been taken of different hair types which is then scaled to fixed normalization and scalability. We are then enhancing the features of datasets by using different deep learning models. Several data augmentation techniques, including rotation, flipping, contrast modifications, zooming, cropping, and Gaussian noise addition, are then used to improve the model's generalization and avoid over fitting. The use of different features are then used to improve the model's flexibility using different clinical conditions. The segmentation techniques used in our model helps to isolate the scalp and hair regions from the background and then removing unwanted elements in the regions and then boosting the feature extraction [10]. To eliminate noise and reflections that could obstruct an accurate diagnosis, sophisticated filtering techniques like median filtering and wavelet denoising are used to get better results. The preparation of data involves ethical implications in addition to technical improvements to enhance the accuracy. To abide by privacy laws like GDPR and HIPAA, all patients scalp disease photos are de-identified to remove personally identifiable information. Specific permission is obtained if the images are collected through different medical records, and steps are taken to ensure dataset diversity by including images from different age groups, genders, and ethnic backgrounds to minimize model biases in the society. The dataset is customized according to the needs for machine learning applications by putting in place a adhere pre-processing pipeline, which enables the trained model to produce precise, trustworthy, and objective predictions for the detection of hair diseases.

Feature Selection: After obtaining the datasets of different images of the diseases the next phase involves the process of feature scaling so that the data would be best scaled to get the desired outputs. In the reach to differentiate between illness of the scalp or a hair sickness detection system's feature selection process involves extracting important qualities from photos, such as their texture, color, and shape or nature. By extracting suitable features from the dataset, Principal Component Analysis (PCA) and deep feature extraction techniques like CNNs are used which can be further enhance the accuracy and efficiency of the developed model. To ensure that only the most informative characteristics are used, dimensionality reduction techniques like Principal Component Analysis (PCA) are applied [11]. PCA aims in reducing the complexity of high-dimensional image data while keeping the essential patterns, hence boosting computational efficiency and minimizing redundant information. Furthermore, we can also use the hierarchical clustering approach to contains the frequently used datasets using different approaches like agglomerative and decisive approaches to get the best desired output to generate results which are more accurate. For more learning of hierarchical representation of hair diseases deep feature extraction techniques are used which is termed as Convolutional Neural Networks (CNNs).

Model Training: For the identification of the different visual patterns Convolutional Neural Networks (CNNs) technology has been used for the hair identification system based on machine learning models or algorithms. Convolutional Neural Networks (CNNs) technology is well suited for assessing the Dermal analysis pictures of hair and their scalp problems as we are using different hierarchical approaches. In order to work with the CNN based model we have to learn complex properties including texture variations, follicular structures, color patterns, and scalp anomalies, the preprocessed dataset must first to be fed into the model. Different techniques we are using like data augmentation, batch normalization, and dropout regularization are used to improve learning efficiency and avoid over fitting from the datasets. Akey factor in increase model performance is hyperparameter adjustment [12]. The hyperparameters may vary according to the need of the models (like epsilon) to increase their accuracy. Other than that techniques like grid search and Bayesian optimization are also used to fine-tune different parameters including learning rate, batch size, number of convolutional layers, kernel size, and activation functions which are then required to improve the efficiency of the dataset. Moreover, pre-trained models like VGG16, ResNet, or Inception—which have previously contained strong feature representations including extensive image datasets—are used to apply transfer learning. We also need to improve the model's capacity by summarize across various hair disease patterns and speed up convergence by fine-tuning these models on our domain-specific dataset. The stability has been improved during training stage using strategies like adaptive learning rate schedulers and early stopping. Other parameters like Accuracy, precision, recall, and F1-score are among the measures used to utilize thoroughly to examine the model's performance, guaranteeing to provide best or effective evaluation of the model's capacity to accurately categorize various hair illnesses. Precision and recall are required in evaluating the trade-off between false positives and false negatives which are then used in confusion metrics for accuracy which offers a general indicator of accurate predictions. A more reliable statistic, particularly when there is a class imbalance, is provided by the F1-score, which is a harmonic mean of precision and recall [13]. The model's classification behavior across many categories is further visualized through the analysis of confusion matrices and ROC-AUC curves. In the need to improve parameters like confidence and transparency in AI-driven dermatological diagnosis and treatments, model interpretability approaches like Grad-CAM (Gradient-weighted Class Activation Mapping) are also used to evaluate significant image regions that contribute to the model's predictions. The model has been developed into a very precise and dependable instrument or architecture for automated hair illness identification through constant training, opening the door for its incorporation into clinical decision-support systems and telemedicine platforms [14].

Model Testing: After the model has been trained the need of the researcher for the deployment of the model is testing. The model testing is a crucial step for accessing the effectiveness and credibility of the machine learning model for hair disease diagnosis. The model's ability to adapt is evaluated using an independent test dataset consisting of previously unseen images. In order to evaluate that the model performance has been evaluated under realistic conditions, this dataset comprises a wide range of hair or scalp conditions or illness including differences in illumination, texture, and scalp features. In order to evaluate the model's performance and generalize across arrange of demographics, the test dataset is carefully selected to contain balanced depictions of various hair conditions in addition to pictures from patients of various ages, genders, and races. Key evaluation metrics like accuracy, precision, recall, and F1-score are utilized to measure the model's performance and they are compared to evaluate the accuracy of the model's performance. Precision is the proportion of anticipated affirmative cases that are really correct, whereas accuracy gives an overall assessment of the correct classifications. In order to reduce false negatives in medical diagnosis, recall gauges measurement of the model's capacity to identify true positive cases [15]. A fair assessment is guaranteed by the F1-score, which is a harmonic mean of precision and recall, especially when working with unbalanced datasets. To assess the model's efficacy in differentiating between various hair illnesses, the Receiver Operating Characteristic (ROC) curve and Area Under the Curve (AUC) analysis are also conducted. Cross-validation methods like k-fold cross-validation are used to further confirm the model's robustness and dependability. By splitting the dataset into several subsets, this technique lowers the possibility of biased findings from unplanned data splits by training and testing the model iteratively on various partitions. Furthermore, real-world testing is done by evaluating the model on clinical images from external sources in order to confirm its adaptability outside of the original dataset. In addition to numerical measurements, a thorough error analysis is carried out to identify cases when the model misdiagnoses certain hair conditions. Misclassifications may result from biases in the dataset, poor image quality, or visual cues that overlap between situations. Techniques like Grad-CAM (Gradient-weighted Class Activation Mapping) are used to visualize which parts of an image influenced the classification result in order to comprehend the model's predictions. This promotes transparency and makes it possible for dermatologists to verify the precision of AI-generated diagnoses. The research paper details testing difficulties, including class imbalances, false positives, false negatives, and variations in Trichoscopic imaging quality. To solve these constraints, potential methods are proposed, such as fine-tuning the model, expanding the dataset, or using new feature extraction techniques. Through a rigorous testing process, the study ensures that the machine learning model is not only accurate but also robust, interpretable, and appropriate for real-world clinical use in the diagnosis of hair disorders [16].

Results: Using images dataset, the machine learning-based model of hair disease identification system has been demonstrated impressive accuracy in correctly diagnosing a variety of hair problems. By applying deep learning techniques like Convolutional Neural Networks (CNNs) and feature extraction methods, the system has shown excellent accuracy, precision, recall, and F1-score in a range of hair diseases detection, including alopecia areata, tinea capitis, and androgenetic alopecia. These results demonstrate the model's ability to identify subtle differences in texture, color, and shape—all of which are critical for an accurate diagnosis. Dermatologists can use the method to aid in early diagnosis, which may lead to more rapid and effective treatment plans. Its effective processing and interpretation of trichoscopic images is the reason behind this [17]. Additionally, a comparison with existing diagnostic methods shows the advantages of using AI-based methods, such as faster image interpretation, reduced diagnostic subjectivity, and increased consistency. Considering the model may be included into clinical decision-support systems to assist medical practitioners in making more precise diagnoses due to its high confidence scores in diagnosing hair illnesses. Additionally, by using dermatologists can better understand which parts of the image influenced the model's decision-making by using explainable AI techniques like Grad-CAM visualizations, which increases trust in AI-driven diagnosis. Nevertheless, additional validation is required to ensure broader application across a range of patient demographics and imaging settings, despite the system's optimistic results. As AI technology advances, automated detection systems could significantly expand access to specialized healthcare by being included into dermatology clinics and telemedicine platforms, particularly in regions with a dearth of dermatological expertise. Figure 12.1 depicts the proposed system's framework, going over each stage in detail, from feature extraction and image pre-processing for classification and model training. This framework serves as a roadmap for developing scalable AI-powered diagnostic tools that can assist dermatologists in giving their patients better care [18].

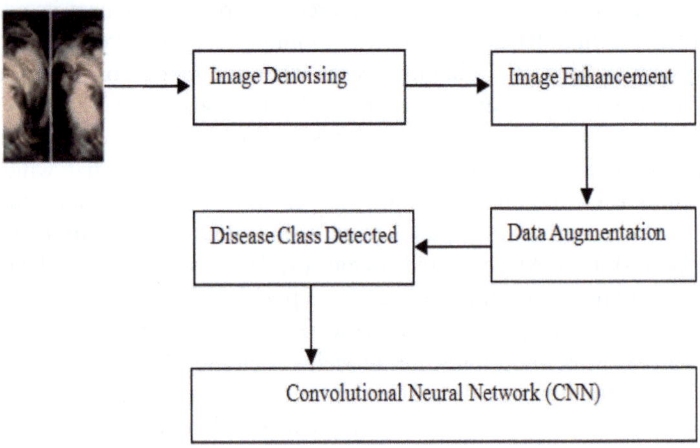

Fig. 12.1 Proposed system framework for hair detection

Below is a detailed explanation of the proposed system framework:

Image Denoising: Enhancing the quality of input images for precise analysis is a crucial preprocessing method known as image denoising. The system efficiently eliminates noise while maintaining important hair-related data by utilizing a range of denoising techniques, including wavelet transform, median filtering, Gaussian blur, and deep learning-based algorithms. The technology use the best possible balance between feature retention and noise reduction by seamlessly integrating denoising into the pipeline and fine-tuning parameters based on noise characteristics. The system is continuously improved and iterated upon based on expert feedback, and advancements in denoising techniques contribute to the system's ability to diagnose hair disorders with greater accuracy and dependability.

Image Enhancement: Image enhancement emerges as a pivotal component in improving the quality and interpretability of input images. By employing machine learning algorithms tailored to the task of image enhancement, we aim to mitigate common challenges such as noise, low contrast, and artifacts that can obscure critical details essential for accurate diagnosis. Through the utilization of Deep Learning models, particularly CNNs, we seek to train robust algorithms capable of learning intricate patterns and features specific to hair diseases from large-scale datasets. By harnessing the power of machine learning, our proposed image enhancement approach not only aims to enhance the visibility of subtle abnormalities associated with various hair conditions but also facilitates more reliable and precise disease classification. Through rigorous experimentation and evaluation, our research endeavors to demonstrate the efficacy and potential of machine learning-driven image enhancement techniques as valuable tools in advancing the field of hair disease detection, ultimately contributing to improved diagnostic accuracy and patient care.

Data Augmentation: In the field of detecting hair illnesses with machine learning, data augmentation emerges as a critical strategy for improving model resilience and performance. Given the scarcity and variability of labeled datasets in this domain, data augmentation techniques provide an excellent method for artificially expanding the training sample pool. The system may generate numerous versions of the original images using augmentation methods such as rotation, flipping, scaling, cropping, and the introduction of random noise, allowing the model to acquire more robust and generalized features. Furthermore, techniques such as elastic deformation and color jittering can replicate fluctuations in scalp and hair conditions, which adds to the dataset [19].

Diseases Class Detection:"Disease class detection" refers to the process of identifying or categorizing different types or groups of diseases based on specific traits or characteristics. In the context of hair disease detection, this involves recognizing and classifying various hair and scalp conditions such as alopecia, dandruff, scalp psoriasis, fungal infections, and more. This approach is crucial in hair disorder detection systems as it allows for the differentiation of various hair conditions, leading to more precise and effective diagnostic and treatment recommendations. By accurately classifying hair disorders, healthcare providers can offer personalized treatment plans tailored to each patient's individual needs. Additionally, disease classification supports the monitoring of disease progression and the effectiveness of treatments over time, leading to better results for individuals affected by hair and scalp conditions. Moreover, it enables tracking of an illness's development and the success of medical interventions, further enhancing patient care. Overall, the categorization of diseases plays a significant role in improving the effectiveness and efficiency of hair disease diagnosis and treatment strategies.

12.5 Results and Discussion

The machine learning model was 90% accurate in recognizing various hair illnesses as shown in Fig. 12.2, with 92% precision as shown in Fig. 12.2, 88% recall as shown in Fig. 12.3, and a 90% F1-score as shown in Fig. 12.3. It successfully detected distinctive patterns for illnesses such as alopecia, dandruff, scalp psoriasis, and fungal infections, which is consistent with medical knowledge.

Cross-validation shown its robustness; however, limitations include dataset availability and interpretability issues. The model has substantial therapeutic implications for early diagnosis and treatment planning. Future research will focus on optimizing architecture, including varied data sources, and addressing ethical considerations for patient privacy and bias reduction.

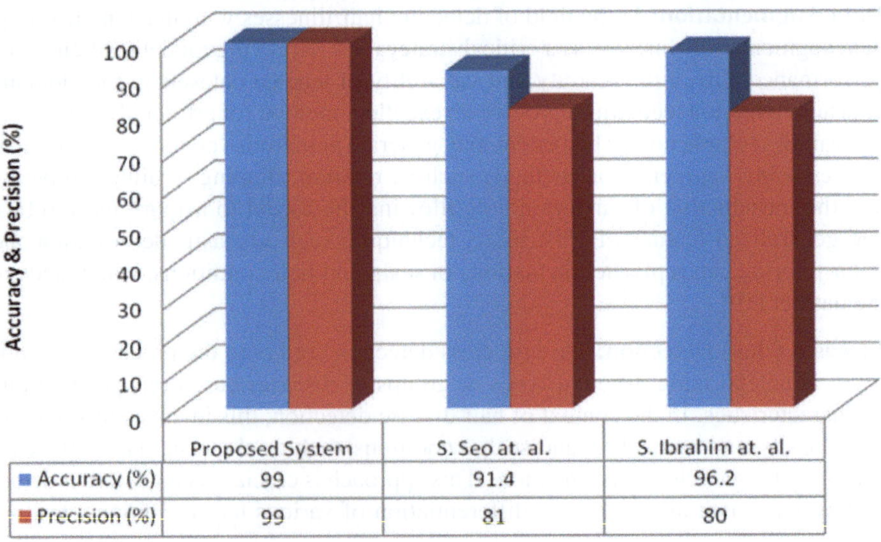

Fig. 12.2 Comparative analysis of proposed with existing system on accuracy and precision

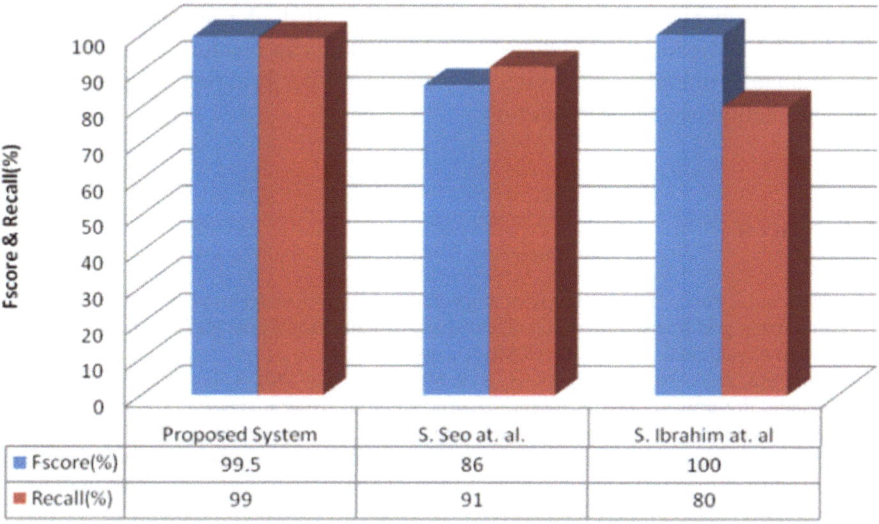

Fig. 12.3 Comparative analysis of proposed with existing system on fscore and recall

12.6 Conclusion

To sum up, our research showed that machine learning methods are both feasible and effective in accurately identifying and categorizing hair issues. Our developed model demonstrated exceptional performance in correctly detecting a range of hair diseases, such as androgenetic alopecia, tinea capitis, and alopecia areata, by utilizing an extensive dataset of dermatoscopic pictures. And also, we are able to attain the great level of accuracy by developing the desired model using different ML based techniques. The best possible feature extraction and feature selection techniques are used to get the greatest rate of accuracy. This states that how machine learning algorithms are greatly used to develop prediction based models to get the early diagnosis to treat the patients. The different.

Visuals which are used in our study are of high resolution which are included in our datasets based on different parameters of image processing using the CNN based techniques. These models would be very helpful to develop other models too to get early or better diagnosis for various other diseases. Furthermore, there is a strong chance to improve the functionality of hair illness detection systems by the incorporation of Convolutional Neural Networks (CNNs), a subset of deep learning architectures. In image processing tasks, CNNs have proven to be more capable than humans, frequently outperforming them in the recognition of subtle visual elements. We will be able to get the increased rates of accuracy through the Deep neural networks. Furthermore, in the upcoming time period, we can also proceed to work on the neura link technology to get the better or more accurate results by using their direct implementation on the patients. Additionally, by using explainable AI techniques, model conclusions may become easier to understand, increasing medical professionals' acceptance and trust. We are getting closer to creating an intelligent, helpful diagnostic system that can revolutionize patient care and enhance dermatological results globally by further bridging the gap between artificial intelligence and dermatology.

References

1. S. Seo and J. Park, "Tracheoscopy of alopecia areata: hair loss feature extraction and computation using grid line selection and eigenvalue," Computational and Mathematical Methods in Medicine, vol. 2020, Article ID 6908018, 9 pages, 2020.
2. S. Ibrahim, Z. A. Noor Azmy, N. N. Abu Mangshor, N. Sabri, A. F. Ahmad Fadzil, and Z. Ahmad, "Pre-trained classification of scalp conditions using image processing," Indonesian Journal of E lectrical E ngineering and Computer Science, vol. 20, no. 1, pp. 138–144, 2020Z.
3. F. Navarro, M. Escudero-Viñolo, and J. Bescós, "Accurate segmentation and registration of skin lesion images to evaluate lesion change," IEEE Journal of Biomedical and Health Informatics, vol. 23, no. 2, pp. 501-508, Mar. 2019.
4. Shabnam Sayyad and Divya Midhunchakkaravathy,"Deep Study on Alopecia Areata Diagnosis for Hair Loss Related Autoimmune Disorder Problem using machine learning".
5. W.-S. Huang, B.-K. Hong, W.-H. Cheng, S.-W. Sun, and K.-L. Hua, "A cloud-based intelligent skin and scalp analysis system," In Proceedings of the 2018 IEEE Visual Communications and Image Processing (VCIP), Taichung, Taiwan, Dec. 2018, pp.
6. S.-H. Lee and C.-S. Yang, "An intelligent hair and scalp analysis system using camera sensors and Norwood-Hamilton model," International Journal of Innovative Computing, Information, and Control, vol. 14, no. 2, pp. 503-518, 2018.
7. W.-J. Chang, M.-C. Chen, L.-B. Chen, Y.-C. Chiu, C.-H. Hsu, Y.-K. Ou, and Q. Chen, "A mobile device-based hairy scalp diagnosis system using deep learning techniques," in Proceedings of the 2020 IEEE 2nd Global Conference on Life Sciences and Technologies (LifeTech), Kyoto, Japan, Mar. 2020, pp. 145–146
8. H. Kim, W. Kim, J. Rew, S. Rho, J. Park, and E. Hwang, "Evaluation of hair and scalp condition based on microscopy image analysis," in Proceedings of the 2017 International Conference on Platform Technology and Service (PlatCon), Busan, Korea, Feb. 2017, pp. 1–4.
9. Hiroyasu T, Hayashinuma K, Ichikawa H, Yagi N. Preprocessing with image denoising and histogram equalization for endoscopy image analysis using texture analysis. Annu Int Conf IEEE Eng Med Biol Soc. 2015; 2015: 789-92.
10. Lalmuanawma S, Hussain J, Chhakchhuak L. Applications of machine learning and artificial intelligence for Covid-19 (SARS-CoV-2) pandemic: A review. Chaos Solitons Fractals. 2020; 139(110059): 110059.
11. M. Roy, A. T. Protity "Hair and Scalp Disease Detection using Machine Learning and Image Processing" published in 2023 European Journal of Information Technologies and Computer Science. (ML) approach. Computer Math Methods Med. 2021; 2021: 1102083.
12. J.-P. Su, L.-B. Chen, C.-H. Hsu, W.-C. Wang, C.-C. Kuo, W.-J. Chang, W.-W. Hu, and D.-H. Lee, "An intelligent scalp inspection and diagnosis system for caring hairy scalp health," in Proc. IEEE 7th Global Conf. Consum. Electron. (GCCE), Nara, Japan, Oct. 2018, pp. 507508.
13. Alshahrani A. A., Al-Tuwaijri R., Abuoliat Z. A., Alyabsi M., AlJasser M. I., Alkhodair R. Prevalence and clinical characteristics of alopecia areata at a tertiary care center in Saudi Arabia. Dermatology Research and Practice.
14. Mrinmoy-Roy. Scalp-Hair-Diseases-Detection. [Internet]. 2022 [cited 2022 October 30]. Available from: https://github.com/MrinmoyRoy/Scalp-Hair-Diseases-Detection.git.
15. Sameer. Image equalization (contrast enhancing) in python - analytics Vidhya – medium. [Internet]. Analytics Vidhya. 2020 [cited 2022 October 29]. Available from: https://medium.com/analyticsvidhya/image-equalization-contrast-enhancing-in-python82600d3b371c.

16. Cotsarelis G. Gene expression profiling gets to the root of human hair follicle stem cells. J Clin Invest. 2006; 116(1): 19-22.
17. Chan CS, Van Voorhees AS, Lebwohl MG, Korman NJ, Young M, Bebo BF Jr, et al. Treatment of severe scalp psoriasis: The Medical Board of the National Psoriasis Foundation. J Am Acad Dermatol. 2009; 60(6): 962-71
18. Virupakshappa, Amarapur B. An automated approach for brain tumor identification using ANN classifier. In: 2017 International Conference on Current Trends in Computer, Electrical, Electronics and Communication (CTCEEC). IEEE; 2017: 1011–6.
19. Kapoor I, Mishra A. Automated classification method for early diagnosis of alopecia using machine learning. Procedia Comput Sci. 2018; 132: 437-43.
20. Balasaravanan K, Prakash M. Detection of dengue disease using artificial neural network based classification technique. Int J Eng Technol. 2017; 7(1.3): 13.

Chapter 13
Blockchain Associated Machine Learning Approach for Earlier Prognosis and Preclusion of Osteoporosis in Elderly

Kottaimalai Ramaraj, Pallikonda Rajasekaran Murugan, Gautam Amiya, Vishnuvarthanan Govindaraj, Muneeswaran Vasudevan, M. Thirumurugan, Yu-Dong Zhang, S. Sheik Abdullah, and Arunprasath Thiyagarajan

Abstract Osteoporosis (OP) also known as porous bone, is a serious illness wherein individual's bones weaken, increasing the likelihood of fractures. OP is caused by micro-architectural degradation of bone tissues, that raises the probability of bone fragility and can result in bone fractures even when no force is placed on it. The estimation of bone mineral density (BMD) is a prevalent method for detecting OP. For women who have reached menopause, prompt and precise forecast and preventative measures of OP is absolutely essential. BMD can be measured using imaging methods like Computed Tomography (CT) and Dual Energy X-ray Absorptiometry (DEXA/DXA). Blockchain (BC) is a revolutionary technique that is utilized in the health sector to store and sharing patient information between clinics, testing centres, dispensaries, and practitioners. The application of blockchain could perhaps reliably detect drastic and even serious errors. As a result, it has the potential to enhance the privacy, and disclosure of clinical data exchange in the health care sector. This system helps health organizations in bringing awareness and improving the evaluation of health records. By integrating the blockchain technology with machine learning algorithms, the various bone disorders such as Osteopenia, Osteoporosis

K. Ramaraj (✉) · P. R. Murugan · M. Vasudevan · S. Sheik Abdullah
Department of Electronics and Communication Engineering, Kalasalingam Academy of Research and Education, Krishnankoil, Tamilnadu, India
e-mail: m.p.raja@klu.ac.in

G. Amiya
Department of Computer Science and Engineering, Kalasalingam Academy of Research and Education, Krishnankoil, Tamilnadu, India

V. Govindaraj · A. Thiyagarajan
Department of Biomedical Engineering, Kalasalingam Academy of Research and Education, Krishnankoil, Tamilnadu, India

M. Thirumurugan
Consultant Orthopaedic Surgeon, MGR Medical University, Chennai, Tamilnadu, India

Y.-D. Zhang
School of Informatics, University of Leicester, Leicester, UK

© The Author(s), under exclusive license to Springer Nature Switzerland AG 2026

K. Dalip et al. (eds.), *Healthcare 5.0*, https://doi.org/10.1007/978-3-032-09582-4_13

and Osteoarthritis can be identified in earlier stage, that delivers a report regarding the prediction of fracture risk. The developed system can assist the physicians and radiologists for more rapid and better diagnosis of the affected ones. In this paper, we propose a fully automated mechanism for suspicious osteoporosis patients that uses machine learning techniques to improve prognosis and preciseness via different processes. Here we developed an automated system that integrates Principal component analysis (PCA) with weighted k-nearest neighbours algorithm (wkNN) to identify, predict and classify the BMD scores as Normal, Osteopenia and Osteoporosis efficaciously. The classified results are validated with the DEXA scan results and by the clinicians to demonstrate the efficacy of the machine learning techniques employed.The laboratories use BC to safely and anonymously share the findings with the patient and doctors.

Keywords Blockchain (BC) Technology · Osteoporosis (OP) · Principal component analysis (PCA) · Weighted k-nearest neighbours algorithm (wkNN) · Bone mineral density (BMD) · Dual Energy X-ray Absorptiometry (DEXA/DXA)

13.1 Introduction

Healthcare records encompass the physical information concerning our bodies and is essential for diagnosing and treating diseases. Medical data consists of numerous archives of patient information that are vital for appropriate treatment and areas for further research. Moreover, it must be kept securely and distributed in order to safeguard the data's confidentiality [1]. With the fast progress of artificial intelligence (AI), healthcare information has emerged as a valuable asset that really can support in the creation of AI diagnostic modelling techniques that could support therapists in the diagnosis. Even though medical data documentation has progressed from paper files to electronic health records, that are more useful for accessing data and retrieval, more consideration should be given to data security. Numerous hospitals and organisations had already whittled down transfer and sharing of data to mitigate personal data breaches, resulting in the development of repository as health record is dispersed within and between diverse healthcare centres [2].

The data sharing operation among the external organizations benefits patients by allowing them to analyse their data using latest methodologies and tools to detect a disorder. Moreover, patients are concerned about the lack of openness in the information exchange procedure because their records could be revealed to a third person. This demonstrates the necessity for a technological system that helps eliminate intermediary entities, reduces expenses, and boosts patient honesty and trust. Blockchain is a technology advancement that may effectively and reliably share data using its decentralised paradigm in order to address this issue [3]. When accessing and evaluating the patient's vital information that is housed across many healthcare systems, a difficult situation arises. By offering the backing of reliable and decentralised

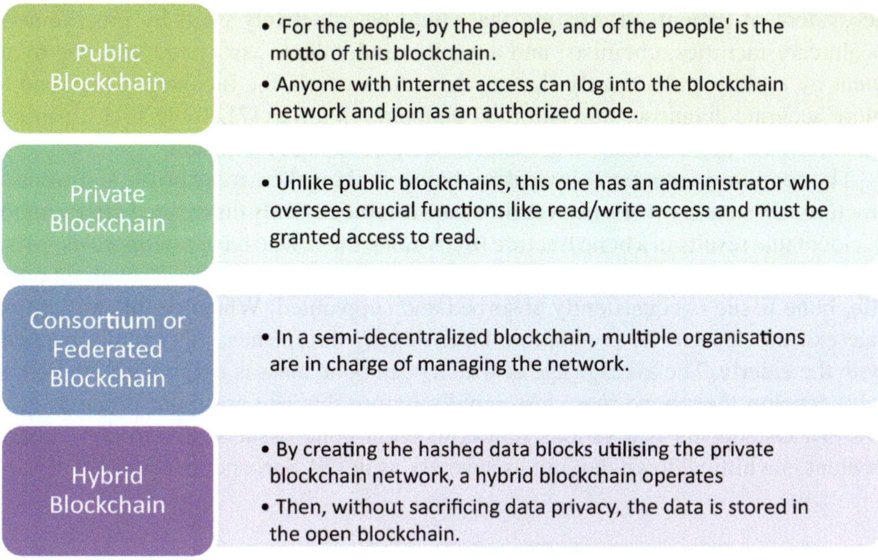

Fig. 13.1 Types of blockchain

databases, blockchain implementation can address this issue with health information exchange platforms. Figure 13.1 depicts the various types of blockchain.

Because of its decentralised and tamper-proof characteristics, Blockchain (BC) technology is extensively utilized in the healthcare industry to manage records. In the medical system, the BC network is employed to store and share patient records among clinics, diagnostic/testing centers, pharmaceutical enterprises, and healthcare professionals [4]. The implementation of BC in the medical system can reliably detect intense and perhaps even serious errors. By delivering immutable data of clinical studies and findings, BC technology could indeed assist in reducing the number of fraudulent activities and errors in clinical study documents. Furthermore, Blockchain-enabled technology has the potential to address outcome transition and specific disclosure in clinical studies. Then, without sacrificing data confidentiality, the data is stored in the open blockchain [5].

Blockchain technology has numerous applications and utilizes in health care services. Distributed ledger technology allows for the secure transmission of patient health records, the management of the drug supply chain, and the unlocking of genetic code by medical research groups [6]. The tendency of blockchain to maintain an immortal, decentralised, and seamless record of all patient information tends to make it an innovation ripe for safety applications. Moreover, whereas blockchain is translucent, it is also secret, trying to conceal any individual's identity with intricated and secure codes capable of protecting the responsiveness of health information. Obtaining access to a patient's health records takes time, which depletes staff resources and slowdowns patient care. Blockchain-based health records provide a solution to these problems. The technology's decentralised nature establishes a single ecosystem of patient information that could be effectively cited by practitioners, healthcare facilities, chemists, and any other individuals associated with the treatment by rapidly and securely sharing data. As a result, the blockchain can lead to more accurate diagnoses and tailored treatment initiatives [7]. Table 13.1 illustrates the numerous applications involved in healthcare using blockchain.

The condition, osteoporosis weakens bones, making them more brittle and prone to fracture. OP steadily worsens over time, and is typically only discovered after a minor accident that results in a bone fracture [8]. Among those who have osteoporosis, wrist, hip, and vertebral fractures are possibly the most frequent wounds. Throughout one's life, bone tissue is consistently absorbed and supplanted. Whenever the absorption rate exceeds the rate of production, bone density diminishes; this usually happens with the elderly. The average age at which peak bone mass is obtained is 20. People who develop their bone mass less rapidly before this age are more prone to have OP [9]. Osteopenia is a serious illness wherein bone tissue's protein and mineral content is whittled down, but not as severely as in OP. Sarcopenia is the age-related

Table 13.1 Applications of Blockchain in Healthcare

BLOCKCHAIN IN HEALTHCARE	Securing patients data
	Streamline care and prevent costly mistakes
	Breakthrough in Genomics
	Point of care Genomics management
	Managing Electronic Medical/Health Record (EMR/EHR)
	Interoperable Electronics Health Records
	Medical supply chain management
	Ease in drug traceability
	Remote patient monitoring
	Personal health record data management
	Diseases and outbreaks being tracked
	Claims for health insurance
	Mobile health Apps
	Easy access and storage of patient information
	Data collection and analysis
	Cybersecurity
	Clinical trials
	Personalized healthcare
	Accelerating Research & Development
	Giving patients control of their data
	Neuroscience studies
	Telemedicine and E-Healthcare

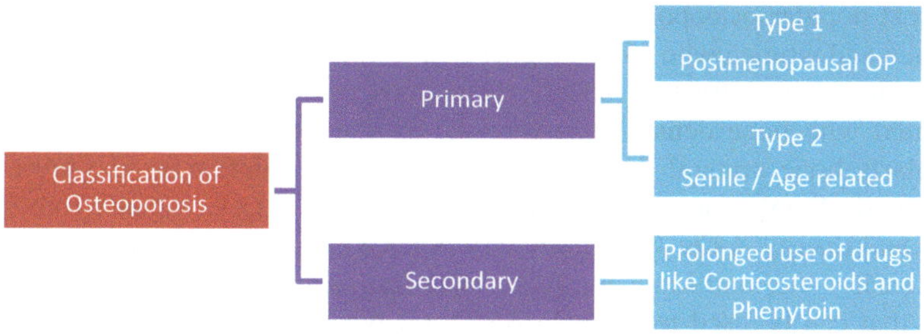

Fig. 13.2 Classification of osteoporosis

deficit of skeletal muscle mass and strength. OP is classified into two categories and is portrayed in Fig. 13.2.

Because the repercussions are only experienced when a fracture occurs, OP is underrated and medicinally silent. Such a subclinical illness is an incredibly complicated problem that affects women all over the world, with postmenopausal women accounting for 80% of cases. It lowers the quality of life for elderly women unless it is addressed or prevented.

Osteoporosis is a prevalent and eerily quiet disorder unless fractures aggravate it. According to estimates, 20% of men and 50% of women over 50 will, at some time in their life, suffer from an osteoporosis-related fracture. The condition occurs primarily in postmenopausal women over the age of 50. Following menopause, bone loss expedites due to a decrease in oestrogen as well as other sex hormone secretion. It is estimated that one-fourth of the world's female population over the age of 60 has OP [10]. Osteoporosis medications can rise bone density, although the gains may seem as mild, they can have a significant impact on fracture rates. Within the first three to four years of treatment, medication can considerably increase bone density in the hip by around 1–3% and in the spine by about 4–8%. Hip fractures and spinal fractures are both reduced by 30–50% and 30–70%, respectively, by the drug. After starting medication, promising benefits may start to show as soon as 6 to 12 months later.

The best defence against osteoporosis is a healthy diet that is rich phosphorus, vitamin D and calcium throughout life, especially in the early postmenopausal years. Vitamin D is made in the skin by exposure to sunlight and is also consumed through diet. Physical activity that is moderate and consistent helps to prevent bone loss. Weight-bearing exercises (weightlifting, walking, and running), Balance exercises, Postural exercises and exercise for strengthening and flexibility should all be a part of physical therapy for those with osteoporosis or even osteopenia [11]. A significant public health issue that millions of older people face is osteoporosis. In addition to leading to fractures, the disease has serious mental and economic ramifications for the sufferer. An interprofessional team of healthcare professionals is most suited to address the disorder because it has numerous risk factors. The Osteoporosis is treated by utilising medicine to strengthen bones and by treating and preventing fractures.

The most popular osteoporosis drugs are bisphosphonates, which are provided to both men and women who have a higher risk of fracture.

Prior to when a fracture occurs, suspicious osteoporosis can be detected or confirmed with bone density examination. Estimation of bone mass and quality, particularly bone mineral density (BMD), are used to determine the presence of osteoporosis. Single Energy X-ray Absorptiometry (SEXA), Computed Tomography (CT), Dual Energy X-ray Absorptiometry (DEXA), Quantitative Ultrasound (QUS) bone densitometer and Magnetic Resonance Imaging (MRI) are some of the techniques that can be used to evaluate BMD. Of which, DEXA is considered as the gold standard technique,by doctors to assess BMD. QUS, which uses sound waves rather than radiation for testing, is yet another commonly used approach to assessing BMD [12]. The metrics Broadband Ultrasound Attenuation (BUA) and Speed of Sound (SoS) are used to calculate the T-score of a patient who has undergone a QUS evaluation.

The most beneficial bone density test is DEXA scan, that uses both high- and low-energy x-rays of the hip and spine two areas where significant fractures are most prone to occur [13]. Doctors can compute bone density using the difference in x-ray findings between low- and high-energy scans. A T-score is given as a result, comparing the subject's bone density to that of a healthy individual of the same gender, race, and ethnicity at the time of peak bone mass, that is around age 30. The T-score decreases when bone density decreases [10]. The Z-score contrasts an individual's bone density with that of a typical person their own age and gender. DXA scans take 10 to 15 min to complete, are painless, and expose the patient to very little radiation. In addition to helping with diagnosis, they might also be helpful for tracking treatment response. The T-score and Z-score range is mentioned in Table 13.2.

With the help of AI and ML, the data can be easily categorised in to three different classes. Initially, the features of three distinct classes are extracted using the method called feature extraction [14]. The process of turning highly dimensional data into a useful representation of lower dimensionality is known as feature extraction. Principal Component Analysis (PCA) is the best feature extraction technique that transforms raw data into processable numerical features that retain the information from

Table 13.2 T-Score and Z-Score rating system

Category	T-Score range	Remarks	Category	Z-Score range	Remarks
Menopausal women and men ≥ 50 years	$+1$ to -1	Healthy bone density	Premenopausal women and men < 50 years	≤ -2	Below the expected range for age
	-1 to -2.5	Osteopenia			
	-2.5 to -3.0	Osteoporosis		> -2	Within the expected range for age
	Less than -3.0	Severe Osteoporosis			

the original data set. Compared to using machine learning on the raw data directly, it produces better outcomes. The representations that are extracted can frequently be used to reduce computational complexity and raise the classifier's accuracy. Once the features are extracted, the data can be clustered based on the similarity and the distance measurement and classified using the neural network classifier. A non-parametric, supervised training classifier called the k-nearest neighbours algorithm (kNN) employs closeness to classify or anticipate how a particular data point will be clustered/grouped. The usefulness of our approach is demonstrated by the experimental findings on three high dimensional data sets. The novel technique that combines PCA and weighted kNN performs well in classifying different classes with high accuracy. The results are stored and safely transferred/shared through the internet to other hospitals or diagnostic centres using the block chain technology employed [15].

The following is how the manuscript is organised: Sect. 13.1 addresses the connotation of BC in healthcare, osteoporosis detection using artificial intelligence andmachine learning, and the way it assists health care professionals in diagnosing and treating bone ailments earlier. Recent researches pertaining to BC and OP are reviewed in Sect. 13.2. The explanation about the datasets utilized in this work is described in Sect. 13.3. The implementation and outcomes of the developed system are explained in Sect. 13.5. The impact of outcomes in healthcare data is concluded in Sect. 13.6.

13.2 Literature Review

With the advent ofmachine learning and artificial intelligence, the ailments or anomalies existing in medical images/datasets can indeed be accurately extracted or classified. The information/data obtained from the outcome's aids in the early prediction and diagnosis of the disease. In recent times, numerous researchers have examined BMD scores collected from CT, X-ray, and DEXA images to diagnose bone disease as osteopenia/osteoporosis [16].

Fang et al. recommended a fully automated system to segment the vertebral body and to assess the BMD score from CT images acquired from 1449 patients [17]. The U-Net was employed for segmentation and DenseNet-121 type DCNN is utilized for evaluating the BMD score. The segmented vertebral body was compared and validated with the manually sketched portion of the vertebral body, which was accepted as ground truth.

Tomita et al. developed an LSTM network to identify and aggregate the radiological features of Osteoporotic vertebral fractures (OVFs) present in the CT scan examined at pelvis, abdomen and chest sites of 1432 images [18]. The incorporation of deep CNN in predicting the ailments in CT has achieved 89.2% accuracy and a 90.8% F1 score. Such a fully automated detection method may shorten the time and manual stress on radiologists for inspection of OVF, and also the possibility of false negatives.

Yamamoto et al. tested five CNN models to predict and assess the osteoporosis from 1131 hip radiography DXA images collected during 2014 to 2019 [19]. Out of five CNN models, the EfficientNet b3 and GoogleNet performed better than other models and attained the best precision, accuracy and specificity values. The clinical covariates such as gender, age, body mass index, fracture history was recorded in this study and the same was included in the ensemble model at the time of detecting the osteoporosis in DXA images.

Using deep learning algorithm, Jang et al. suggested a model to predict the osteoporosis from DXA hip images of 1001 patients [20]. Out of 1001 images, 800 were allotted for training, 100 for validation and remaining 100 for testing. Based on VGG16 equipped with non-local neural network, the DNN was utilized and it achieved 81.2% accuracy, 91.1% sensitivity, and 68.9% specificity.

An automated system for the screening of OP from Dental Panoramic Radiographs (DPRs) by examining the BMD using DCNN was introduced by Lee et al. To assess the effect of various transfer learning strategies on DL, four different study groups were utilized [21]. By employing gradient-weighted class activation mapping method, the visual representation of the best features are identified and noted in the mandibular region.

Shim et al. developed an automated model to predict OP in postmenopausal Korean women [22]. The backward stepwise variable selection technique was employed for selecting the features. Seven different neural network classifier models were utilized in this study for assessing OP values from continuous and categorical based values. On the whole, the ANN model perfectly predicts the OP in the dataset and achieved best area under the region of curve value than others.

Nakamoto et al. formulated a system to predict the OP by identifying the low bone mass in dental panoramic X-ray images, femoral neck and lumbar spine using DCNN [23]. The average accuracy attained in predicting OP in femoral neck and lumbar spine was 76% and 73%, respectively.

Zhang et al. established a model for classifying OP and osteopenia from X-ray images of 1616 patients examined at lumbar spine [24]. T-score values derived from DXA images were used as reference for the classification. Using DCNN the classification of normal data and low bone bass data were obtained.

Many researchers have attempted to predict OP in CT and X-ray images, but only very few tried on DEXA images. Traditional approaches, in particular CNN, provide better classification accuracy when looking for OP in multimodal images. However, there is scope for further improvement by using some other variation ofa neural network classifier to increase classification accuracy while minimizing error rates. To estimate and categorize three different classes of data from the entire dataset, we here proposed weighted k Nearest Neighbour (kNN) method. One of the most used techniques for data classification and clustering is kNN. In tests on various datasets, it was discovered that the kNN algorithm operated quite efficiently. To achieve high classification accuracy with less computing time, the weight factor is combined with traditional kNN.

13.3 Methodology

13.3.1 Principal Component Analysis (PCA)

Large datasets are becoming more common in many fields of study. These records need to have their dimensionality significantly reduced in order to be interpreted, while still retaining the majority of the data's information. Principal component analysis (PCA), one of the earliest and perhaps most extensively utilised unsupervised learning algorithms, has been developed specifically for this purpose among many other techniques [25]. With the aid of orthogonal transformation, it is an empirical method that transforms the measurements of correlated features into a sequence of linearly uncorrelated characteristicsknown as the Principal Components. This tool is most commonly used for explorative data analysis and predictive modelling [26].

The number of these PCs is either equal to or less than the original features present in the dataset.

Initially, the covariance matrix is computed by PCA determines the covariance matrix's eigen vectors and eigen values. The data are then projected along the eigen vectors. If the original data is n-dimensional, it is possible to reduce it to k-dimensionality, where $k \leq n$.

Following are the steps involved in PCA technique.

1. Standardize the range of continuous initial variables

In this procedure, the range of variables is determined and standardised to examine the impact of each variable equitably. The variables that are influencing the other variables in short range can be categorised with the aid of the initial variables' computation. As an outcome, the analysis' final results will be more subjective. By utilizing the formulas given below, the variable transformation can be achieved.

$$\mathrm{z} = \frac{Value - \mu}{\sigma} \tag{13.1}$$

$$\mu = \frac{Sum\ of\ the\ terms}{Total\ number\ of\ terms} \tag{13.2}$$

$$\sigma = \sqrt{\frac{\sum (x_i - \mu)^2}{N}} \tag{13.3}$$

where,

z = Standardized values.

μ = Mean.

σ = Standard Deviation of dataset.

x_i = Each value from the dataset.

N = Number of values in the dataset.

2. Computation of covariance matrix

The covariance matrix helps in identifying the values that are not highly related to the dataset, by contrasting it with the mean value.

A 2-D covariance matrix is given as

$$Covariance\,Matrix = \begin{bmatrix} COV(X,X) & COV(X,Y) \\ COV(Y,X) & COV(Y,Y) \end{bmatrix} \tag{13.4}$$

where,

$$Covariance = \frac{Sum(X - \mu_x)(Y - \mu_y)}{Number\ of\ data\ points} \tag{13.5}$$

The covariance of a number with itself is its variance like,

$$COV(X,X) = Var(X) \text{ and } COV(Y,Y) = Var(Y) \tag{13.6}$$

The Covariance Matrix's entries will also be symmetric with regard to the fact that the covariance is commutative at the diagonal elements.

$$COV(X,Y) = COV(Y,X) \tag{13.7}$$

The data points are said to be correlated if the covariance matrix value is positive. The negative value of the matrix represents the data points are inversely correlated.

At last, it can be identified as which pair of variables are correlated with each other, and that will help in categorization.

3. Calculate the eigenvalues and eigenvectors of the covariance matrix to find the principal components

Let a square matrix is represented as A. The non-zero vector v, is an eigenvector of A if

$$Av = \lambda v \tag{13.8}$$

where λ=eigenvalue

4. Form a feature vector to finalize theprincipal components
5. Reorganize the data along the principal component axes

$$Final\ dataset(X_{ij}) = Standardized\ original\ dataset * Feature\ vector \tag{13.9}$$

13.3.2 Weighted kNN

An easy-to-use supervised machine learning approach is the k-nearest neighbours (KNN) algorithmhas the potential to overcome the classification and regression issues [27]. A supervised learning system produces the required output after receiving new unlabelled data and uses labelled input data to train a function. Based on the distance measurements, it modifies the training data and classifies the recently received test data. It locates the test data's k-nearest neighbours, and the majority of the class labels then carry out the classification. For a data analyst, choosing the best K value to achieve the highest accuracy is an ongoing challenge.

A revamped version of the KNN algorithm is termed weighted KNN. The selection of the hyperparameter k is one of several factors that influence how well the KNN model performs [28]. The classifier would become more susceptible to outlier data points if k were chosen as too minimal. The neighbourhood may retain an excessive number of points from other classes if k is too sizable. Weighted kNN is employed to circumvent this drawback. The closest k points are given a weight in weighted kNN. The idea behind a weighted KNN is to give points that are closer together more weight and less to the points that are farther apart.

The basic function that is employed is the inverse distance function, which implies that weight significantly reduces as distance level increases and rises as distance decreases.

The steps to be followed in kNN are mentioned below.

1. Load the unclassified training and testing data (X_{ij})—mentioned in Eq. (13.9).
2. Compute the distance from the new data to all others data that is previously classified using Euclidian, Minkowski, Manhattan, or Weighted distance method.
3. Choose the value of k, by taking the square root of the total number of data points.
4. Compute the distance between test and training data in each row by anyone of the distance measuring methods.
5. Sort the values in ascending order as per their distance value
6. Choose top k rows from the sorted values in array.
7. A class type is assigned to the test point based on the most frequent class of these rows.
8. Classifies the new data with the class.

13.3.3 Proposed PCA-WkNN

The integration of neural network approach with feature extraction technique aids in identifying and classifying the disease precisely and with better accuracy [29, 30]. In this study, we combine PCA with weighted kNN, which not only reduces data dimensionality to expedite kNN computation and moreover, it reduces redundant data while maintaining useful information to enhance kNN predictive accuracy. The distinctive features noticed through PCA are then provided for classification. The data was divided into testing and training for the classification task. By using supervised learning, 30% of the data were used for testing and 70% for training.

Figure 13.3 demonstrates the proposed block diagram that employs blockchain in healthcare.

13.4 Dataset Description

Obtaining the vast quantity of BMD score is very difficult because most hospitals in Tamilnadu are not equipped with DEXA equipment. The BMD score of patients from DEXA images was obtained from the Tamil Nadu Government Multi Super Speciality Hospital, Omandurar Estate, Chennai, Tamilnadu, after a drawn-out process on request through the appropriate procedures. In this dataset, BMD scores of patients with ages ranging from 25 to 74 were presented. DEXA images are taken at a spine location from all patients. The clinicians classified the patient's bone as normal, osteopenic, or osteoporosis based on the BMD score value attained.

The equipment used to collect the information was the 64-detector Hologic Discovery Wi DXA bone densitometry system. High-definition Instant Vertebral Assessment, which creates a high-resolution image for assessing vertebral fractures. The usual exposure time for diagnosing femur and lumbar spine bone disorders is 10 s/0.04 mGy. The following are the DXA equipment operational requirements: 60° to 90° F, 100 VAC (16 A), 20–80% relative humidity, and 3,400 BTU/hr average heat load.

13.5 Implementation and Results

The baseline characteristics of three distinct groups are exhibited in Table 13.3: men, pre- and postmenopausal women. The baseline characteristics include the number of subjects, age and menopause information for women.

Among the 100 subjects including young and elders, 37 cases (37%) had normal/typical bone mass, 42 cases (42%) had decreased bone mass namely osteopenia, and 21 cases (21%) had osteoporosis. The considerable ratio between normal bone mass and bone mass loss decreased significantly with age. As per the BMD Score values obtained from DEXA scan images, Table 13.4 lists the number of subjects at various ages who are all classified as normal, osteopenia, and osteoporosis.

The number of men in different age groups is categorized into three classes based on BMD Score values and is exemplified in Table 13.5. Out of 49 men from overall subjects, 19 men (38.77%) had normal bone mass, 21 men (42.85%) had reduced bone mass/osteopenic and 9 men (18.36%) had severe bone mass loss/osteoporotic.

The number of Premenopausal women in different age groups is categorized as normal, Osteopenic, and Osteoporotic based on BMD Score values and is represented

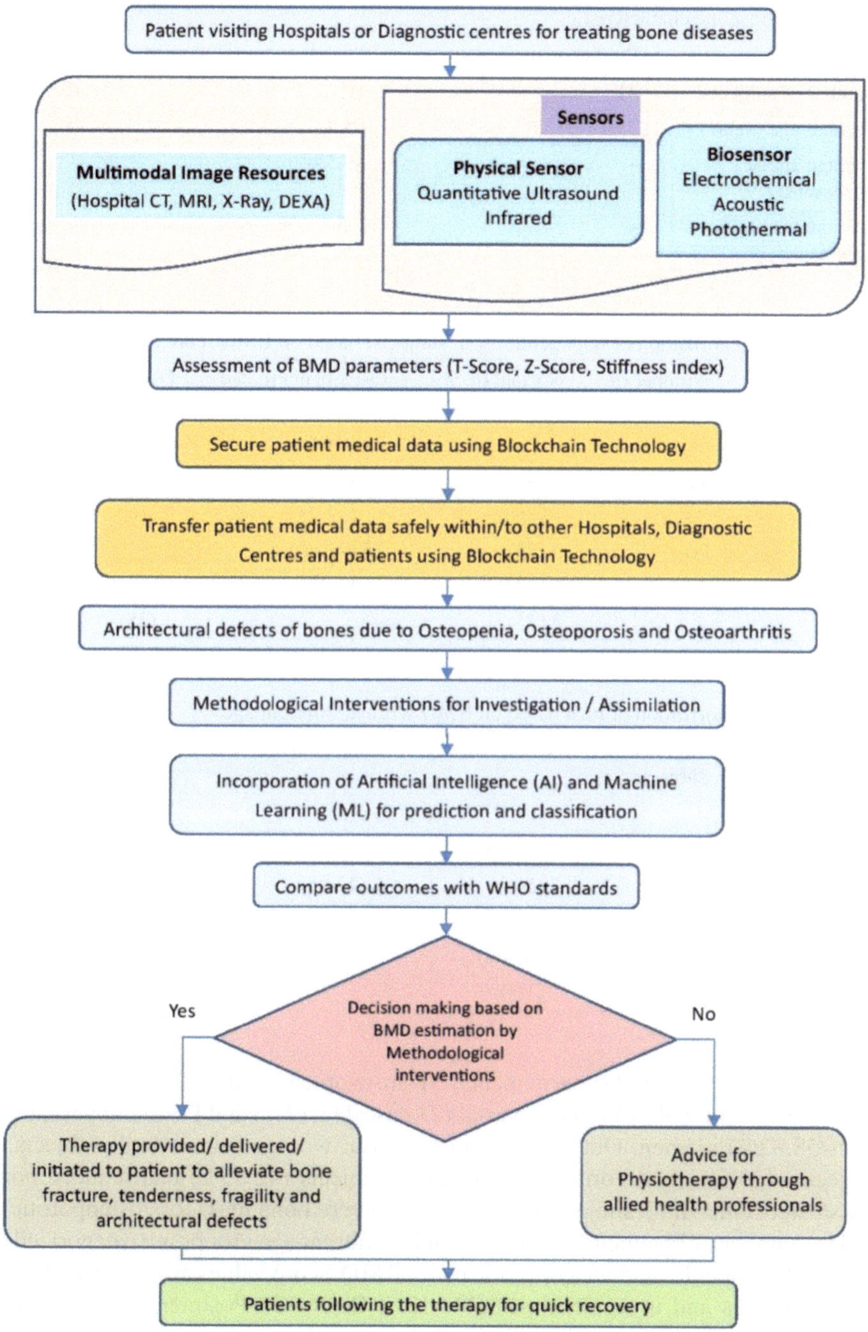

Fig. 13.3 Proposed block diagram employing blockchain in healthcare

Table 13.3 Baseline characteristics

Characteristics	Premenopausal women	Postmenopausal women	Men
Number of subjects	41	10	49
Average age (years)	35.4	55.4	49.2
Average age at menopause	–	49.5	–
Average years since menopause	–	7.5	–

Table 13.4 Categorization of both genders (all subjects) based on BMD score at different ages

Age range (in years)	Number of subjects	Normal (n, %)	Osteopenia (n, %)	Osteoporosis (n, %)
21–30	6	–	4 (66.66)	2 (33.33)
31–40	12	5 (41.66)	6 (50)	1 (8.33)
41–50	28	14 (50)	11 (39.28)	3 (10.71)
51–60	36	9 (25)	15 (41.66)	12 (33.33)
61–70	14	6 (42.85)	6 (42.85)	2 (14.28)
> 70	4	3 (75)	–	1 (25)

Table 13.5 Categorization of men based on BMD score at different ages

Age range (in years)	Number of subjects	Normal (n, %)	Osteopenia (n, %)	Osteoporosis (n, %)
21–30	4	–	3 (75)	1 (25)
31–40	7	3 (42.85)	3 (42.85)	1 (14.28)
41–50	14	7 (50)	6 (42.85)	1 (7.14)
51–60	16	3 (18.75)	8 (50)	5 (31.25)
61–70	6	4 (66.66)	1 (16.66)	1 (16.66)
>70	2	2 (100)	–	–

in Table 13.6. Among 100 participants, 51 are women. Again, the women participants are categorized as Premenopausal (21 cases/41.17%) and Postmenopausal (30 cases/58.82%) women. Out of 21 Premenopausal women from overall subjects, 9 subjects (42.85%) had normal bone mass, 9 subjects (42.85%) had reduced bone mass/osteopenic and 3 subjects (14.28%) had severe bone mass loss/osteoporotic.

The number of Postmenopausal women in different age groups are categorized as normal, Osteopenic, and Osteoporotic using BMD Score values accomplished from DEXA images and is signified in Table 13.7. Out of 30 Postmenopausal women from overall subjects, 9 subject (30%) had normal bone mass, 12 subjects (40%) had reduced bone mass/osteopenic and 9 subjects (30%) had severe bone mass loss/osteoporotic.

Table 13.6 Categorization of premenopausal women based on BMD score at different ages

Age range (in years)	Number of subjects	Normal (n, %)	Osteopenia (n, %)	Osteoporosis (n, %)
21–30	2	–	1 (50)	1 (50)
31–40	5	2 (40)	3 (60)	–
41–50	14	7 (50)	5 (35.71)	2 (14.28)

Table 13.7 Categorization of Postmenopausal women based on BMD score at different ages

Age range (in years)	Number of subjects	Normal (n, %)	Osteopenia (n, %)	Osteoporosis (n, %)
51–60	20	6 (30)	7 (35)	7 (35)
61–70	8	2 (25)	5 (62.5)	1 (12.5)
> 70	2	1 (50)	0	1 (50)

The overall subjects are clustered as men, Pre- and Postmenopausal women. The assessment of normal, osteopenic and osteoporotic bone mass is evaluated using the BMD scores attained and is consolidated in Table 13.8. In accordance with the research, bone deterioration occurs with age in both men and women, but it happens more quickly in postmenopausal women.

The BMD values measured using DEXA, the gender and age details of the subjects are given as input to the proposed PCA-wkNN classifier for identifying and categorizing the three groups of subjects and is exemplified in Table 13.9.

The proposed classifier identifies the bone ailment by utilizing the BMD values in the feature extraction stage,and as an outcome, the subjects were classified using wkNN and are remarked as normal, osteopenia or osteoporosis. The classesare identified and is illustrated in Table 13.10.

The disease identified subjects are then grouped and categorized into three distinct classes for a better introspection and meaningful representation, and is depicted in Table 13.11.

The final GUI representation of the proposed algorithm, and its interactions involved in prediction of the three types of disease classes is made available in

Table 13.8 Assessment of osteoporosis on all subjects based on BMD score

Group	Number of subjects	Normal (n, %)	Osteopenia (n, %)	Osteoporosis (n, %)
Premenopausal women	21	9 (42.85)	9 (42.85)	3 (14.28)
Postmenopausal women	30	9 (30)	12 (40)	9 (30)
Men	49	19 (38.77)	21 (42.85)	9 (18.36)

Table 13.9 BMD scores with patient age details given as input for classification to the algorithm proposed

Patient no	Gender	Age	T-Score	Patient no	Gender	Age	T-Score
1	M	29	−1.7	51	M	56	1.200
2	M	69	−0.3	52	F	63	0.721
3	M	49	−1.3	53	M	44	0.949
4	M	35	0	54	F	54	0.730
5	F	62	−1.8	55	F	55	0.937
6	F	54	−1.0	56	M	39	0.868
7	F	55	−2.7	57	M	73	1.111
8	F	52	−4.0	58	F	63	0.917
9	F	73	17.7	59	M	58	0.850
10	F	51	−2.9	60	F	65	0.979
11	M	28	−1.8	61	M	60	0.892
12	F	35	−0.6	62	F	49	0.947
13	M	43	−3.1	63	F	49	0.821
14	F	22	−2.9	64	M	41	1.0
15	F	62	−1.7	65	F	49	−3.0
16	F	58	−2.9	66	F	49	−1.3
17	M	55	0.7	67	F	33	−2.9
18	M	30	−2.6	68	M	75	−1.0
19	M	34	−0.6	69	F	45	−2.0
20	F	42	−1.0	70	M	50	0.2
21	M	50	−0.2	71	M	57	−1.2
22	M	52	−2.3	72	M	64	−2.2
23	F	61	−1.5	73	F	44	−0.6
24	F	52	−1.8	74	F	51	−1.8
25	M	33	−1.5	75	M	42	−0.9
26	F	67	−0.7	76	M	31	−2.1
27	F	61	−1.2	77	M	52	−0.7
28	F	71	−2.9	78	F	29	−1.8
29	M	55	−2.8	79	F	37	−0.8
30	F	51	−1.1	80	M	46	−1.5
31	M	49	−1.7	81	M	43	−0.7
32	F	33	−1.5	82	M	53	−1.8
33	M	55	−4.3	83	M	51	−0.2
34	F	47	−2.6	84	F	58	−2.6
35	F	48	−1.1	85	F	56	−2.6
36	M	58	−1.4	86	M	34	1.4

(continued)

Table 13.9 (continued)

Patient no	Gender	Age	T-Score	Patient no	Gender	Age	T-Score
37	M	44	−1.7	87	F	51	0
38	M	60	−2.6	88	F	54	−1.3
39	M	48	−2	89	F	49	−3.3
40	M	34	−1.2	90	M	61	−3.2
41	M	62	−1.0	91	M	56	−2.0
42	F	59	−2.5	92	M	57	−1.5
43	F	50	−0.1	93	M	21	−0.5
44	F	60	−1.8	94	F	48	0.5
45	F	59	−2.9	95	F	52	−1.8
46	F	56	−1.1	96	M	57	−1.3
47	M	43	−0.4	97	M	44	0.7
48	F	34	−0.8	98	M	68	−2.4
49	M	65	−0.5	99	F	50	−0.6
50	F	50	−1.9	100	F	51	−1.0

Fig. 13.4 to have a better visual perception/picturization, and for future development/improvement of the same.

The true positive, true negative, false positive and false negative values attained while classifying the BMD score of patients into three different classes are expressed as a confusion matrix in Fig. 13.5. Out of 100 patients, the misclassified data were up to 9, which gives the overall success rate in the prediction of the data to be as 91%. The same is detailed in the confusion matrix accordingly.

13.6 Conclusion

Using the BC utilized in this work, the consumer/customer can independently verify themselves in order to retrieve the health records among all healthcare professionals. The BC's authentication mechanism, which creates trustworthy setup helps the patients to restrict/manage the volume of information shared with a third party. Such an aspect of BC works well with the patient-centered medical system because it fosters trust and advances technology. The data that has been tapped onto the blockchain is immutable; no one else can add to, subtract from, or modify it.

Table 13.10 BMD scores classification done using wkNN for the nature/type of disease identification

Patient no	Gender	T-Score	Remarks	Patient no	Gender	T-Score	Remarks
1	M	−1.7	Osteopenia	51	M	1.0	Normal
2	M	−0.3	Normal	52	F	−3.0	Osteoporosis
3	M	−1.3	Osteopenia	53	M	−1.3	Osteopenia
4	M	0	Normal	54	F	−2.9	Osteoporosis
5	F	−1.8	Osteopenia	55	F	−1.0	Normal
6	F	−1.0	Normal	56	M	−2.0	Osteopenia
7	F	−2.7	Osteoporosis	57	M	0.2	Normal
8	F	−4.0	Osteoporosis	58	F	−1.2	Osteopenia
9	F	17.7	Normal	59	M	−2.2	Osteopenia
10	F	−2.9	Osteoporosis	60	F	−0.6	Normal
11	M	−1.8	Osteopenia	61	M	−1.8	Osteopenia
12	F	−0.6	Normal	62	F	−0.9	Osteopenia
13	M	−3.1	Osteoporosis	63	F	−2.1	Osteopenia
14	F	−2.9	Osteoporosis	64	M	−0.7	Normal
15	F	−1.7	Osteopenia	65	F	−1.8	Osteopenia
16	F	−2.9	Osteoporosis	66	F	−0.8	Normal
17	M	0.7	Normal	67	F	−1.5	Osteopenia
18	M	−2.6	Osteoporosis	68	M	−0.7	Normal
19	M	−0.6	Normal	69	F	−1.8	Osteopenia
20	F	−1.0	Osteopenia	70	M	−0.2	Normal
21	M	−0.2	Normal	71	M	−2.6	Osteoporosis
22	M	−2.3	Osteoporosis	72	M	−2.6	Osteoporosis
23	F	−1.5	Osteopenia	73	F	1.4	Normal
24	F	−1.8	Osteopenia	74	F	0	Normal
25	M	−1.5	Osteopenia	75	M	−1.3	Osteopenia
26	F	−0.7	Normal	76	M	−3.3	Osteoporosis
27	F	−1.2	Osteopenia	77	M	−3.2	Osteoporosis
28	F	−2.9	Osteoporosis	78	F	−2.0	Osteopenia
29	M	−2.8	Osteoporosis	79	F	−1.5	Osteopenia
30	F	−1.1	Normal	80	M	−0.5	Normal
31	M	−1.7	Osteopenia	81	M	0.5	Normal
32	F	−1.5	Osteopenia	82	M	−1.8	Osteopenia
33	M	-4.3	Osteoporosis	83	M	−1.3	Osteopenia
34	F	−2.6	Osteopenia	84	F	0.7	Normal
35	F	−1.1	Osteopenia	85	F	−2.4	Osteoporosis

(continued)

Table 13.10 (continued)

Patient no	Gender	T-Score	Remarks	Patient no	Gender	T-Score	Remarks
36	M	−1.4	Osteopenia	86	M	−0.6	Normal
37	M	−1.7	Osteopenia	87	F	−1.0	Normal
38	M	−2.6	Osteoporosis	88	F	1.2	Normal
39	M	−2	Osteopenia	89	F	−2.8	Osteoporosis
40	M	−1.2	Osteopenia	90	M	−1.5	Osteopenia
41	M	−1.0	Normal	91	M	−0.9	Osteopenia
42	F	−2.5	Osteopenia	92	M	−1.1	Osteopenia
43	F	−0.1	Normal	93	M	1.3	Osteopenia
44	F	−1.8	Osteopenia	94	F	0	Normal
45	F	−2.9	Osteoporosis	95	F	−1.7	Osteopenia
46	F	−1.1	Normal	96	M	−1.3	Osteopenia
47	M	−0.4	Normal	97	M	−0.7	Normal
48	F	−0.8	Normal	98	M	−0.1	Normal
49	M	−0.5	Normal	99	F	−0.6	Normal
50	F	−1.9	Osteopenia	100	F	−1.2	Osteopenia

This important feature of BC offers the attribution and authenticity which is necessary to help stop the drug counterfeiting and forgery. The PCA-based weighted KNN algorithm is proposed in this research to address some of the drawbacks of the conventional KNN algorithm. The intended outcomes are achieved, which shortens the KNN algorithm's time complexity. The fusion of this BC allows for the safe storage and transmission of medical records of patients pertaining to OP to certain other clinics and diagnostic centres. The incorporation of AI and ML aids in early detection, prediction, and diagnosis of OP by utilizing the BMD parameters.

Table 13.11 Patient grouping based on the classification of the nature of disease

Patient no	Gender	T-Score	Remarks	Patient no	Gender	T-Score	Remarks
2	M	−0.3	Normal	36	M	−1.4	Osteopenia
4	M	0	Normal	37	M	−1.7	Osteopenia
6	F	−1	Normal	39	M	−2	Osteopenia
9	F	17.7	Normal	40	M	−1.2	Osteopenia
12	F	−0.6	Normal	42	F	−2.5	Osteopenia
17	M	0.7	Normal	44	F	−1.8	Osteopenia
19	M	−0.6	Normal	50	F	−1.9	Osteopenia
21	M	−0.2	Normal	53	M	−1.3	Osteopenia
26	F	−0.7	Normal	56	M	−2	Osteopenia
30	F	−1.1	Normal	58	F	−1.2	Osteopenia
41	M	−1	Normal	59	M	−2.2	Osteopenia
43	F	−0.1	Normal	61	M	−1.8	Osteopenia
46	F	−1.1	Normal	62	F	−0.9	Osteopenia
47	M	−0.4	Normal	63	F	−2.1	Osteopenia
48	F	−0.8	Normal	65	F	−1.8	Osteopenia
49	M	−0.5	Normal	67	F	−1.5	Osteopenia
51	M	1	Normal	69	F	−1.8	Osteopenia
55	F	−1	Normal	75	M	−1.3	Osteopenia
57	M	0.2	Normal	78	F	−2	Osteopenia
60	F	−0.6	Normal	79	F	−1.5	Osteopenia
64	M	−0.7	Normal	82	M	−1.8	Osteopenia
66	F	−0.8	Normal	83	M	−1.3	Osteopenia
68	M	−0.7	Normal	90	M	−1.5	Osteopenia
70	M	−0.2	Normal	91	M	−0.9	Osteopenia
73	F	1.4	Normal	92	M	−1.1	Osteopenia
74	F	0	Normal	93	M	1.3	Osteopenia
80	M	−0.5	Normal	95	F	−1.7	Osteopenia
81	M	0.5	Normal	96	M	−1.3	Osteopenia
84	F	0.7	Normal	100	F	−1.2	Osteopenia
86	M	−0.6	Normal	7	F	−2.7	Osteoporosis
87	F	−1	Normal	8	F	−4	Osteoporosis
88	F	1.2	Normal	10	F	−2.9	Osteoporosis
94	F	0	Normal	13	M	−3.1	Osteoporosis
97	M	−0.7	Normal	14	F	−2.9	Osteoporosis
98	M	−0.1	Normal	16	F	−2.9	Osteoporosis
99	F	−0.6	Normal	18	M	−2.6	Osteoporosis

(continued)

Table 13.11 (continued)

Patient no	Gender	T-Score	Remarks	Patient no	Gender	T-Score	Remarks
1	M	−1.7	Osteopenia	22	M	−2.3	Osteoporosis
3	M	−1.3	Osteopenia	28	F	−2.9	Osteoporosis
5	F	−1.8	Osteopenia	29	M	−2.8	Osteoporosis
11	M	−1.8	Osteopenia	33	M	−4.3	Osteoporosis
15	F	−1.7	Osteopenia	38	M	−2.6	Osteoporosis
20	F	−1	Osteopenia	45	F	−2.9	Osteoporosis
23	F	−1.5	Osteopenia	52	F	−3	Osteoporosis
24	F	−1.8	Osteopenia	54	F	−2.9	Osteoporosis
25	M	−1.5	Osteopenia	71	M	−2.6	Osteoporosis
27	F	−1.2	Osteopenia	72	M	−2.6	Osteoporosis
31	M	−1.7	Osteopenia	76	M	−3.3	Osteoporosis
32	F	−1.5	Osteopenia	77	M	−3.2	Osteoporosis
34	F	−2.6	Osteopenia	85	F	−2.4	Osteoporosis
35	F	−1.1	Osteopenia	89	F	−2.8	Osteoporosis

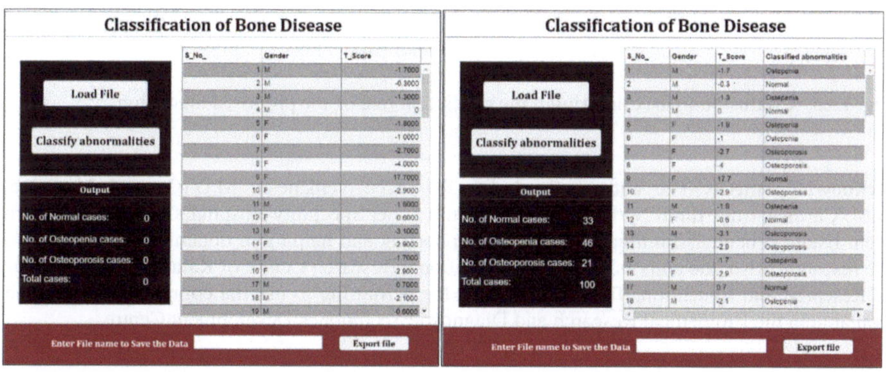

Fig. 13.4 GUI representing the classification of patients based on disease using the proposed algorithm **a** Input GUI **b** Output GUI

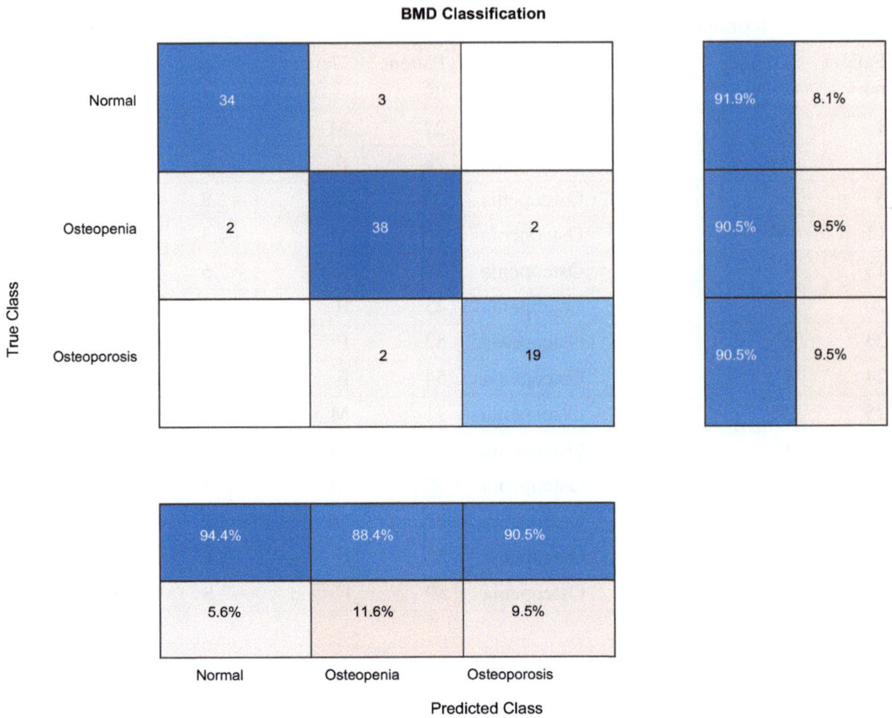

Fig. 13.5 Confusion matrix describing the prediction rate of the proposed algorithm

Acknowledgements The authors thank Dr. R. Sayee Venkatesh, M.D.(General Medicine), D.M.(Cardio), from Tamil Nadu Government Multi Super Speciality Hospital, Omandurar Government Estate, Chennai, Tamilnadu, India, for supporting and helping in obtaining DEXA data for the research. Also, the authors thank the International Research Centre of Kalasalingam Academy of Research and Education, Tamil Nadu, India for permitting to use the computational facilities available in the Biomedical Research and Diagnostic Techniques Development Centre.

References

1. Kumar, M., Dubey, H. and Kumar, K., 2022. Blockchain in Healthcare: A Smart and Hierarchical Decentralized Tracing Application in Healthcare by Blockchain. In *Handbook of Green Computing and Blockchain Technologies*, CRC Press, pp. 133–140).
2. Ali, O., Jaradat, A., Kulakli, A. and Abuhalimeh, A., 2021. A comparative study: Blockchain technology utilization benefits, challenges and functionalities. *IEEE Access*, *9*, pp.12730–12749.
3. Hasan, M.R., Deng, S., Sultana, N. and Hossain, M.Z., 2021. The applicability of blockchain technology in healthcare contexts to contain COVID-19 challenges. *Library Hi Tech*, *39*(3), pp.814–833.

4. Haleem, A., Javaid, M., Singh, R.P., Suman, R. and Rab, S., 2021. Blockchain technology applications in healthcare: An overview. *International Journal of Intelligent Networks, 2*, pp.130–139.
5. Yaqoob, I., Salah, K., Jayaraman, R. and Al-Hammadi, Y., 2022. Blockchain for healthcare data management: opportunities, challenges, and future recommendations. *Neural Computing and Applications, 34*(14), pp.11475–11490.
6. Chen, H.S., Jarrell, J.T., Carpenter, K.A., Cohen, D.S. and Huang, X., 2019. Blockchain in healthcare: a patient-centered model. *Biomedical journal of scientific & technical research, 20*(3), p.15017.
7. Xi, P., Zhang, X., Wang, L., Liu, W. and Peng, S., 2022. A review of Blockchain-based secure sharing of healthcare data. *Applied Sciences, 12*(15), p.7912.
8. Lorentzon, M., Johansson, H., Harvey, N.C., Liu, E., Vandenput, L., McCloskey, E.V. and Kanis, J.A., 2022. Osteoporosis and fractures in women: the burden of disease. *Climacteric, 25*(1), pp.4–10.
9. Moshayedi, S., Tasorian, B. and Almasi-Hashiani, A., 2022. The prevalence of osteoporosis in rheumatoid arthritis patient: a systematic review and meta-analysis. *Scientific reports, 12*(1), pp.1–11.
10. Anabtawi, A., Holyoak, M., He, J., Cristiano, E., Polineni, D. and Graves, L., 2022. Trabecular bone score in people with cystic fibrosis. *Osteoporosis International, 33*(5), pp.1137–1145.
11. Reid, I.R. and Billington, E.O., 2022. Drug therapy for osteoporosis in older adults. *The Lancet, 399*(10329), pp.1080–1092.
12. Kadri, A., Binkley, N., Hernando, D. and Anderson, P.A., 2022. Opportunistic Use of Lumbar Magnetic Resonance Imaging for Osteoporosis Screening. *Osteoporosis International, 33*(4), pp.861–869.
13. Periasamy, K., Periasamy, S., Velayutham, S., Zhang, Z., Ahmed, S.T. and Jayapalan, A., 2022. A proactive model to predict osteoporosis: An artificial immune system approach. *Expert Systems, 39*(4), p.e12708.
14. Amiya, G., Ramaraj, K., Murugan, P.R., Govindaraj, V., Vasudevan, M. and Thiyagarajan, A., 2022. A Review on Automated Algorithms Used for Osteoporosis Diagnosis. *Inventive Systems and Control*, pp.247–262.
15. Smets, J., Shevroja, E., Hügle, T., Leslie, W.D. and Hans, D., 2021. Machine learning solutions for osteoporosis—a review. *Journal of Bone and Mineral Research, 36*(5), pp.833–851.
16. Pedoia, V., Caliva, F., Kazakia, G., Burghardt, A.J. and Majumdar, S., 2021. Augmenting Osteoporosis Imaging with Machine Learning. *Current Osteoporosis Reports*, pp.1–11.
17. Fang, Y., Li, W., Chen, X., Chen, K., Kang, H., Yu, P., Zhang, R., Liao, J., Hong, G. and Li, S., 2021. Opportunistic osteoporosis screening in multi-detector CT images using deep convolutional neural networks. *European Radiology, 31*(4), pp.1831–1842.
18. Tomita, N., Cheung, Y.Y. and Hassanpour, S., 2018. Deep neural networks for automatic detection of osteoporotic vertebral fractures on CT scans. *Computers in biology and medicine, 98*, pp.8–15.
19. Yamamoto, N., Sukegawa, S., Kitamura, A., Goto, R., Noda, T., Nakano, K., Takabatake, K., Kawai, H., Nagatsuka, H., Kawasaki, K. and Furuki, Y., 2020. Deep learning for osteoporosis classification using hip radiographs and patient clinical covariates. *Biomolecules, 10*(11), p.1534.
20. Jang, R., Choi, J.H., Kim, N., Chang, J.S., Yoon, P.W. and Kim, C.H., 2021. Prediction of osteoporosis from simple hip radiography using deep learning algorithm. *Scientific reports, 11*(1), pp.1–9.
21. Lee, K.S., Jung, S.K., Ryu, J.J., Shin, S.W. and Choi, J., 2020. Evaluation of transfer learning with deep convolutional neural networks for screening osteoporosis in dental panoramic radiographs. *Journal of clinical medicine, 9*(2), p.392.
22. Shim, J.G., Kim, D.W., Ryu, K.H., Cho, E.A., Ahn, J.H., Kim, J.I. and Lee, S.H., 2020. Application of machine learning approaches for osteoporosis risk prediction in postmenopausal women. *Archives of osteoporosis, 15*(1), pp.1-9.

23. Nakamoto, T., Taguchi, A. and Kakimoto, N., 2022. Osteoporosis screening support system from panoramic radiographs using deep learning by convolutional neural network. *Dentomaxillofacial Radiology*, *51*(6), p.20220135.
24. Zhang, B., Yu, K., Ning, Z., Wang, K., Dong, Y., Liu, X., Liu, S., Wang, J., Zhu, C., Yu, Q. and Duan, Y., 2020. Deep learning of lumbar spine X-ray for osteopenia and osteoporosis screening: A multicenter retrospective cohort study. *Bone*, *140*, p.115561.
25. Tsuchida, W., Kobayashi, Y., Inoue, K., Horie, M., Yoshihara, K. and Ooie, T., 2022. Kinematic characteristics during gait in frail older women identified by principal component analysis. *Scientific reports*, *12*(1), pp.1–12.
26. Świetlicka, I., Kuniszyk-Jóźkowiak, W. and Świetlicki, M., 2022. Artificial neural networks combined with the principal component analysis for non-fluent speech recognition. *Sensors*, *22*(1), p.321.
27. Fasihi, L., Tartibian, B., Eslami, R. and Fasihi, H., 2022. Artificial intelligence used to diagnose osteoporosis from risk factors in clinical data and proposing sports protocols. *Scientific Reports*, *12*(1), pp.1–11.
28. Ganaie, M.A., Tanveer, M. and Alzheimer's Disease Neuroimaging Initiative, 2022. KNN weighted reduced universum twin SVM for class imbalance learning. *Knowledge-Based Systems*, *245*, p.108578.
29. Thaha, M.M., Kumar, K., Murugan, B.S., Dhanasekeran, S., Vijayakarthick, P. and Selvi, A.S., 2019. Brain tumor segmentation using convolutional neural networks in MRI images. Journal of medical systems, 43(9), pp.1–10.
30. Shankar, K., Elhoseny, M., Chelvi, E.D., Lakshmanaprabu, S.K. and Wu, W., 2018. An efficient optimal key based chaos function for medical image security. IEEE Access, 6, pp.77145–77154.

Chapter 14
GA Versus PSO: Effectiveness in Feature Selection for Heart Disease Prediction

Savita Wadhawan, Sri kant, Loveleena Mukhija, and Amanpreet Kaur

Abstract The World Health Organization attributes 31% of the world population mortality statistics to heart-related diseases making it one of the deadliest diseases in the world. Even though many heart conditions are curable, a timely and precise diagnosis significantly raises the chances of survival. However, procedures like blood tests, ECG, and angiography are incredibly costly and depend heavily on human precision. This study explores how feature selection impacts the accuracy of machine learning (ML) classifiers in diagnosing heart disease. Specifically, it evaluates the effectiveness of two widely used optimization techniques, Genetic Algorithms (GA) and Particle Swarm Optimization (PSO), in selecting the most relevant features. These methods are tested alongside various classifiers, including Support Vector Machine (SVM), K-Nearest Neighbors (KNN), Decision Tree (DT), Naïve Bayes (NB), Random Forest (RF), and Multi-Layer Perceptron (MLP). The research is based on datasets from the UCI repository, which include data from Cleveland, Hungarian, Switzerland, and Long-Beach-Va. Various performance metrics were used in the evaluation of the model including accuracy, sensitivity, specificity, and F-score. The results showed that with Accuracy and computational complexity, the GA faired slightly better or in some cases roughly on par with the PSO. The contribution of the study focuses on the integration of intelligent decision support systems with feature selection methods.

Keywords Genetic algorithm · Particle swarm optimization · Machine learning · Feature selection · Heart disease

S. Wadhawan (✉) · L. Mukhija · A. Kaur
Chitkara University Institute of Engineering and Technology (CUIET), Chitkara University, Rajpura, Punjab, India
e-mail: savitawadhawan@gmail.com

L. Mukhija
e-mail: loveleena.mukhija@chitkara.edu.in

A. Kaur
e-mail: amanpreet.1094@chitkara.edu.in

Sri kant
Department of Academic Affairs, Chandigarh University, Chandigarh, India
e-mail: kaushiksrikant@gmail.com

© The Author(s), under exclusive license to Springer Nature Switzerland AG 2026
K. Dalip et al. (eds.), *Healthcare 5.0*, https://doi.org/10.1007/978-3-032-09582-4_14

14.1 Introduction

Heart disease is a leading cause of death worldwide, as recognized by WHO. According to statistics, approximately 17.9 million deaths are due to heart disease, which is approximately 31% of the total deaths, out of which 85% are due to heart attacks and strokes (https://www.who.int/health-topics/cardiovascular-diseases/#tab=tab1). Myocardial Ischemia, a common cause of irregular heartbeat, chest pain, fatigue, and shortness of breathis the primary reason for heart attacks and strokes. Early detection of these symptoms is essential for timely diagnosis and significantly increases patient lifespan. Timely diagnosis of these symptoms depends upon various tests such as angiography, ECG, ECO, and blood tests, and the experts interpret the results accurately [1]. However, it is quite difficult and costly for a patient to obtain these laboratory tests and consult an expert, which can lead to errors. Therefore, there has been a lot of research interest in developing intelligent decision support systems to automate the heart diagnostic process and reduce human errors.Intelligent decision support systems can be built using various machine learning techniques like support vector machine (SVM), decision tree (DT), k-nearest neighbor (k-NN), etc. [2, 3]. However, the success of these ML techniques highly depends on the quality of data used, especiallythe data pre-processing process. Feature selection is one of the crucial steps of data pre-processing, especially in medical datasets where the data dimensionality highly influences the performance of the systems in terms of accuracy, and system complexity and improves the system generalization by preventing overfitting. Further, feature selection algorithms are categorized into two categories, that is, filter and wrapper methods. On one side, filter methods rely on the statistical/ probabilistic measure and are independent of the performance of the classifier [4]. These methods assign a score to each feature based on the statistical measure and remove the feature having a low score. Some of the common filter methods are Info Gain, Chi-Square, relief etc. Whereas, wrapper methods use the classifier performance to select the best features. Particle Swarm Optimization, Genetic Algorithms (GA), Sine Cosine algorithm (SCA), Grey Wolf optimization (GWO), Whale Optimization algorithm (WOA), and other meta-heuristic algorithms [5, 6] and their hybrid forms are widely used as wrapper methods for feature selection. A successful feature selection requires a global search strategy like evolutionary computation (EC).

This paper accesses the performance of well-known wrapper methods that as genetic algorithm (GA) and particle swarm optimization (PSO) for heart disease classification with various ML classifiers: logistic regression(LR), Naïve Bayes (NB),decision tree (DT), support vector machine (SVM), k-Nearest Neighbour (KNN) and multilayer perceptron (MLP). We consider the heart disease datasets from UCI repositories such as Cleveland (UD_CL), Hungarian (UD_HN), Switzerland (UD_SW), and Long-Beach-Va (UD_Va) datasets for experimentation. The performance metrics considered were accuracy, misclassification rate, true positive rate (sensitivity), true negative rate (specificity), false positive rate, precision, and F-score. This study aims to identify the most effective feature selection method to enhance the predictive accuracy of machine learning classifiers for heart disease classification.

The rest of the paper is organized as follows: The related work is discussed in Sect. 14.2. Section 14.3 presents the method and methodology, including the dataset, considered classifiers, and optimization algorithms. Section 14.4 describes the experimental setup with detailed results and discussion followed by the conclusion in Sect. 14.5.

14.2 Related Work

The author in [7]introduces a new model called binary genetic swarm optimization (BGSO) for feature selection (FS) in pattern recognition models. The proposed BGSO combines the exploitation capability of genetic algorithms with the exploration ability of particle swarm optimization. BGSO refines the information extracted from feature subsets with sequential one-point flipping. The algorithm was tested on 20 datasets from UCI repository to evaluate the algorithm's performance. Two other metaheuristic algorithms, the S-shaped binary Sine Cosine Algorithm (SBSCA) and the V-shaped binary Sine Cosine Algorithm (VBSCA) were proposed in [8] for feature selection. These algorithms generate binary position vectors for each solution in continuous search space. Both BSCA variants were tested on medical datasets showing their efficacy as compared to other algorithms. Another study [9] proposed hybrid classifiers utilizing ensemble models and majority voting to improve the prediction accuracy. The study utilizes a genetic algorithm for data pre-processing and feature selection. The overfitting was handled with tenfold cross-validation. Further, the model was tested on cardiovascular datasets with 98.18% classification accuracy.

The author [10] introduced an adaptive genetic algorithm with a fuzzy logic model with a rough set theory, for heart disease prediction with reduced death risk and treatment cost for developing countries. The author evaluates the model with datasets available on UCI repository. The study [11] introduces a genetic algorithm and weighted fuzzy rule-based CDSS for heart disease prediction which selects the most appropriate feature subset using stochastic inquiry from the genetic algorithm. The proposed model enhances risk forecasting and ensures better outcomes in terms of accuracy, specificity, and sensitivity. On another side, [12] introduced an Adaptive Neuro-Fuzzy Inference System, which utilizesa genetic algorithm to optimize membership functions for heart disease prediction and achieve 91.25% accuracy. The author in [13] utilizes the binary version of the genetic algorithm for feature reduction. The proposed methodology was applied tothe Flavia dataset and got a subset of 100 features. In the paper [14], the author used PCA and hybrid GA to reduce the dimensionality of the heart disease dataset from UCI repository. The proposed methodology achieved 94.06% accuracy in predicting heart disease, avoiding the drawbacks of heuristic clustering methods. Whereas, the author [15] introduced an MLP-PSO hybrid algorithm for heart disease prediction. The author compares the performance of the proposed method with the other ten ML-basedalgorithms for Cleveland heart disease dataset and achieves an accuracy of 84.61%.Another approach GAPSO-RF, which is the hybridization of GA and PSO was introduced in [16] to improve the heart disease prediction. The proposed approach used a multivariate statistical analysis and discriminate mutation strategies to modify GA forthe global search and PSO for the local search, thus achieving an accuracy of 95.6% and 91.4% for the Cleveland and Statlog datasets respectively.

The author in [17] introduced a novel methodology QPSO-SVM to analyze and predict heart disease. The proposed approach utilizes the quantum PSO and SVM for heart disease prediction and achieved an accuracy of 96.31% on the Cleveland dataset. In [18], the author utilizesXGBoost and RF classification methods to analyze heart disease. Also, the study found that the PSO enhances the AUC of heart disease prediction and contributes to more accurate and faster diagnosis of heart disease.

The study [19] used PSO to enhance the performance of various ML classifiers to detect, predict, and classify heart disease, achieving an accuracy of 89.09% with the K-NN classifier. On the other hand, the author in [20] evaluates the efficiency of binary PSO and genetic algorithm to determine coronary artery disease with exercise stress testing data. The study found that BPSO outperforms with less complexity and classification time. Further, the author in [21] used PSO with a boosting approach to extract the rules to recognizecoronary artery disease. The proposed approach reduces the weight of training examples and focuses on fuzzy rules in the next cycle to detect the CAD with acceptable accuracy. Another study [22] used wrapper-based feature selection consisting of a DT with C4.5 as a training program and binary PSO with a mutation operator for spam detection. The proposed approach outperforms other meta-heuristics in terms of classification accuracy without compromising specificity and sensitivity for a dataset consisting of 6000 emails from 2012.

14.3 Method and Methodology

14.3.1 Dataset Used

In this study, we consider the most popular heart disease datasets, that is, Cleveland (UD_CL), Hungarian (UD_HN), Switzerland (UD_SW), and Long-Beach-Va (UD_Va) from the UCI repository for experimentation [23]. In these datasets, we have a total of seventy-six features, out of which we consider only fourteen for experimentation because other features are either not related to the target attribute or may have replicated information. The target attribute HD is based upon the angiographic status with 5 values as 0 for Healthy, 1–4 are diagnosed with stage 1- stage 4. Here, we consider the value 0 as a healthy person and stage 1–4 as the person sufferingfrom heart disease. Originally, we had 303 records in the UD_CL dataset, out of which 164 records hadlabel 0 and 139 records having label 1 that suffered from heart disease. The 294 records of UD_HN datasets consist of 188 records with label 0 and 106 records with label 1. On another side, out of 123 records of the UD_SW dataset, only 8 records have the label 0, and 115 records havethe label 1. Whereas, in UD_Va dataset, we have 200 records in which 51 records are having label 0 and 149 have label 1.

14.3.2 Classifiers Used

Support Vector Machine (SVM)

The SVM classifier creates hyperplane as a decision boundary between data points to classify them into different classes. The hyperplane dimensions are determined by the number of features and different classes are assigned to various data points on either side of these hyperplanes. The hyperplanes with maximum margin, which shows the distance between data points will be considered as given in Fig. 14.1. For binary classification, the instances are classified as $a.x + b = 0$, where (.) represents the dot product. After that, this hyperplane is used to create predictions. The following equations are used to describe the hyperplanes:

$$a.x + b = +1$$

$$a.x + b = -1$$

where $+1$ indicates the points that are above or on the hyperplane, and -1 refers to points that are below the hyperplane.

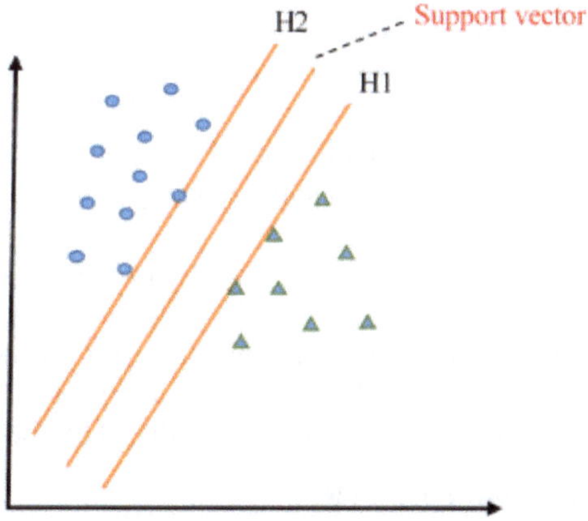

Fig. 14.1 Support vector separating hyperplane between two classes

K-nearest Neighbour (KNN)

The k-nearest neighbor (kNN) is an instance-based supervised learning method. It makes an instance observation based upon the similarity computation of a point to its input sample in the training set by calculating the Euclidian distance(dt) (Eq. 14.1) between an observed point a(1, y1) and a datapoint b(2, y2) from the training set ϵ D^T.

$$dt(A, B) = \sqrt{(x1 - x2)^2 - (y1 - y2)^2} \tag{14.1}$$

After calculating the distance, it identifies k nearest neighbors based upon dt($_{,i}$). The odd value of k yields better for binary classification with smooth decision boundaries.

Decision Tree (DT)

The decision tree is a supervised technique that generates a tree-like structure with decision and leaf nodes. A decision node has two or more branches, with the leaf node at the end of each branch representing the classification or judgment. The decision tree is built with top-down greedy search, starting from the root node to the leaf node. The Gini index is used to randomly identify the piece of information. The lower Gini index will be preferred and is calculated as Eq. 14.2.

$$gini_{index} = 1 - \sum_i p_i^2 \tag{14.2}$$

where probability p_i of an object belonging to a specific class. A 0 value of the Gini index that all records at the node are from the same class, whereas a Gini index of 1 indicates that all records are from different classes.

Naïve Bayes (NB)

The Naïve Bayes (NB) method is a supervised learning algorithm that classifies new feature vectors based on conditional probability. It uses a training dataset to compute the conditional probability of each feature vector for different classes. Once these probabilities are determined, the algorithm assigns a class to a new vector by evaluating the likelihood of it belonging to each class. The probability of a class given a feature is calculated using Eq. 14.3.

$$P(A|b) = (P(A) \cdot P(b|A))/P(b) \tag{14.3}$$

where P(A|b) represents the posterior probability of class A given feature b, P(A) is the prior probability of class A, P(b) is the prior probability of feature b(where, P(b) $\neq$ 0), and P(b|A) representing the likelihood of feature b occurringin class A.

Random Forest (RF)

This classifier enhances its performance by utilizing the decision tree bagging technique. It randomly selects multiple subsets of data from the original dataset through

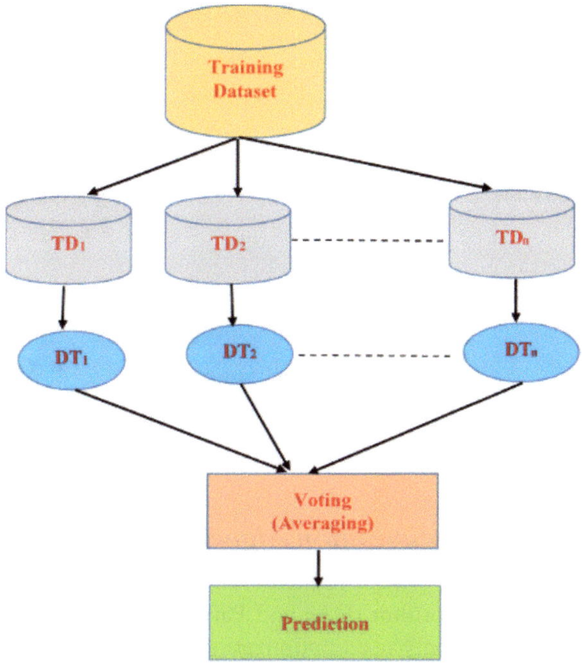

Fig. 14.2 Random forest algorithm

a process called bootstrapping, which allows for sampling with replacement. Each decision tree in the model is trained on a different subset, ensuring diversity among the trees. Once all the trees have made their predictions, a majority voting system is applied to combine their outputs and determine the final classification, as illustrated in Fig. 14.2. RF requires less hyper-parameter adjustment than other related ensembles. Tree mutuality is produced by the original bagging decision tree, which suffers from the consequence of large variation. As a result, RF reduces variance by adding additional randomization to the tree-generation technique.

Multilayer Perceptron (MLP)

A Multi-Layer Perceptron (MLP) is a type of feedforward neural network made up of multiple layers, including an input layer, one or more hidden layers, and an output layer. It is trained using the backpropagation algorithm, which adjusts the weights of the network to minimize errors. As shown in Fig. 14.3, MLP is widely used for both linear and non-linear classification tasks, making it one of the most popular neural network models in machine learning.

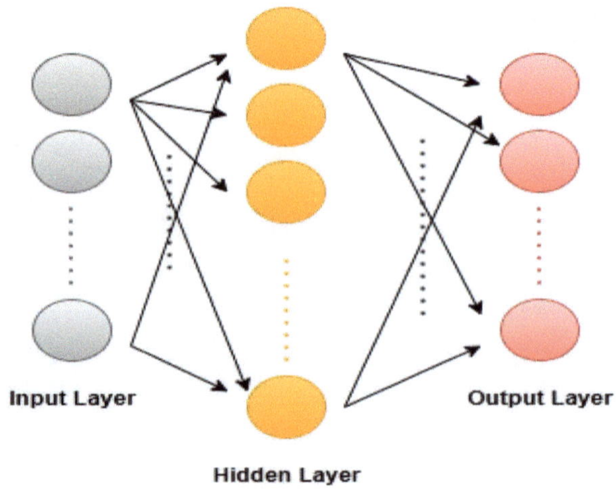

Fig. 14.3 Neural network architecture

14.3.3 Optimization Algorithms

Genetic Algorithm (GA) and Particle Swarm Optimization (PSO) are among the most commonly used optimization techniques [7]. Their performance is evaluated by comparing the experimental results in heart disease classification. To ensure a fair comparison, both algorithms are implemented with the same initial search ranges, population sizes, iteration counts, and objective functions during the optimization process.

Genetic Algorithm: GA is swarm swarm-based metaheuristic optimization technique used to produce optimal solutions for search problems. This algorithm is inspired by the natural behavior of evolution andgenerates the solutions inspired by the methods of inheritance, crossover, mutation, and selection.

How GA Works: Initially, a set of random solutions are generated called populations and the individual solution is called a candidate solution. The population in each iteration is called the generation. Each generation's candidate solution is evaluated through a fitness function for the fitness value. After evaluating the fitness value, a set of candidates is selected with preferred selection criteria for further evolution and modified with some natural process of selection, crossover, and mutation. Then, some of the best solutions are selected randomly to create a new population from the current population. This new generation is used in the next iteration of the algorithm. This process repeats till the stopping criteria are not met. The Fig. 14.4 depicts the framework.

Particle Swarm Optimization (PSO): PSO is a problem-solving technique created by Kennedy and Eberhart in 1995 [24]. It is inspired by how birds fly in groups to search for food. Initially, it wasdeveloped to simulate this natural behavior, further, the researchers found it to be highly effective for solving optimization problems.

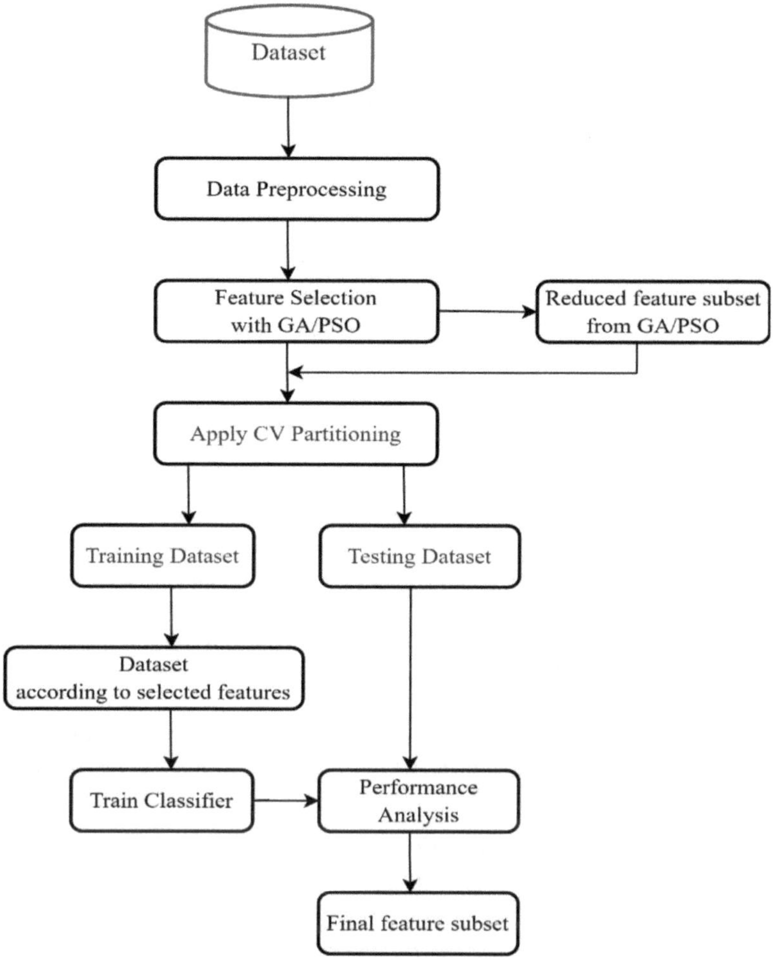

Fig. 14.4 Framework

How PSO Works: PSO works by imitating the social patterns of a flock of birds. It starts with the randomly generated group of possible solutions, called "particles". These particles move within a defined space to find the best possible solution. Each particle has two key properties: its position and velocity, which change as it moves through the search space. This algorithm works to optimize a specific function, called the cost or fitness function, by either minimizing or maximizing its value.Each particle remembers its best solution in terms of particle best(pbest) found so far and the best solution discovered by any particle in the swarm known as global best(gbest) is shared among all particles. During each step of the process: Each particle adjusts its velocity and position based on its own best experience i.e. pbest. It also considers the swarm's overall best position (gbest). This process continues until the best solution is found or the maximum number of iterations is reached.

Advantages of PSO Over Genetic Algorithm (GA)

1. GA requires complex operations like crossover, mutation, and selection, while PSO only needs a few adjustable parameters, making it easier to implement.
2. PSO is less computationally expensive compared to GA, which requires more resources.
3. PSO evaluates both local and global solutions, whereas GA focuses only on the current fitness value.
4. PSO can remember past fitness values, but GA does not have this feature.

14.3.4 Stopping Criteria

The algorithm stops based on specific conditions, such as reaching the maximum number of iterations, evaluating the fitness function, or observing repeated values for the particle's best and global best positions and repeating fitness value, etc. If none of these criteria are met, the global best position in PSO and the candidate with the best fitness in the current population are reported as the optimal feature set.

1. **Experimentation and Result**

The simulations of the proposed methodologywere carried out in Matlab 2019 and executed on a laptop having an intel(R) CORE(TM) i5 7200U @ 2.70 GHz processor with 8 GB RAM and 250 MB SSD.

Performance Evaluation with Genetic Algorithm: Tables 14.1, 14.2, 14.3, and 14.4 represents the performance of the genetic algorithm in terms of average accuracy of 10 runs with 100 iterations each for considered datasets, that is, Cleveland, Hungarian, Switzerland, and Log-Beach-Va datasets respectively. From Table 14.1, it is found that the genetic algorithm gives a better accuracy of 84.158% with the SVM classifier for the Cleveland dataset. For the Hungarian dataset, the genetic algorithm gives better accuracy of 82.993% with the RF classifier (Table 14.2). Table 14.3 shows, the genetic algorithm performs better with an accuracy of 93.496% with SVM and KNN classifiers both for the Switzerland dataset. Whereas, for the Long-Beach-Va dataset, the genetic algorithm gives an accuracy of 75.5% with the RF classifier (Table 14.4).

Performance Evaluation with Genetic Algorithm: Tables 14.5, 14.6, 14.7 and 14.8 represents the performance of PSO in terms of average accuracy of 10 runs with 100 iterations each for considered datasets. From Table 14.5, it is found that PSO gives a better accuracy of 84.488% with SVM and KNN classifiers both for the Cleveland dataset. For the Hungarian dataset, the PSO algorithm gives a better accuracy of 82.993% with the KNN classifier (Table 14.6). Table 14.7 shows, that the PSO performs better with an accuracy of 93.496% with SVM and KNN classifiers

Table 14.1 Performance evaluation of cleveland dataset (UD_CL) with GA

Technique	Accuracy	Misclassification rate	TP rate/ sensitivity	TN rate/ specificity	FP rate	Precision	F-Score
GA + DT	78.878	21.122	71.942	84.756	15.244	80	75.758
GA + RF	83.168	16.832	77.698	87.805	12.195	84.375	80.899
GA + NB	77.558	22.442	69.784	84.146	15.854	78.862	74.046
GA + SVM	**84.158**	15.842	75.54	91.463	8.537	88.235	81.395
GA + KNN	83.498	16.502	79.856	86.585	13.415	83.459	81.618
GA + MLP	82.508	17.492	75.54	88.415	11.585	84.677	79.848

Table 14.2 Performance evaluation of hungarian (UD_HN) dataset with GA

Technique	Accuracy	Misclassification rate	TP rate/ sensitivity	TN rate/ specificity	FP rate	Precision	F-Score
GA + DT	78.231	21.769	66.981	84.574	15.426	71	68.932
GA + RF	**82.993**	17.007	70.755	89.894	10.106	79.787	75
GA + NB	80.952	19.048	73.585	85.106	14.894	73.585	73.585
GA + SVM	81.633	18.367	67.925	89.362	10.638	78.261	72.727
GA + KNN	81.973	18.027	60.377	94.149	5.851	85.333	70.718
GA + MLP	81.633	18.367	59.434	94.149	5.851	85.135	70

Table 14.3 Performance evaluation of switzerland (UD_SW) dataset with GA

Technique	Accuracy	Misclassification rate	TP rate/ sensitivity	TN rate/ specificity	FP rate	Precision	F-Score
GA + DT	88.618	11.382	94.783	0	100	93.162	93.966
GA + RF	88.618	11.382	94.783	0	100	93.162	93.966
GA + NB	80.488	19.512	85.217	12.5	87.5	93.333	89.091
GA + SVM	**93.496**	6.504	100	0	100	93.496	96.639
GA + KNN	**93.496**	6.504	100	0	100	93.496	96.639
GA + MLP	91.87	8.13	98.261	0	100	93.388	95.763

Table 14.4 Performance evaluation of Long-Beach-Va(UD_Va) dataset with GA

Technique	Accuracy	Misclassification rate	TP rate/ sensitivity	TN rate/ specificity	FP rate	Precision	F-Score
GA + DT	72.5	27.5	88.591	25.49	74.51	77.647	82.759
GA + RF	**75.5**	24.5	93.96	21.569	78.431	77.778	85.106
GA + NB	73.5	26.5	95.973	7.843	92.157	75.263	84.366
GA + SVM	74.5	25.5	100	0	100	74.5	85.387
GA + KNN	74	26	97.315	5.882	94.118	75.13	84.795
GA + MLP	74.5	25.5	98.658	3.922	96.078	75	85.217

both for the Switzerland dataset. Whereas, for the Long-Beach-Va dataset, the PSO algorithm gives an accuracy of 77% with the KNN classifier (Table 14.8).

Figure 14.5 illustrates the average number of features selected by the Genetic Algorithm (GA) across different datasets and classifiers. In the Cleveland dataset, GA selects an average of 7.2 features when using the SVM classifier. For the Hungarian dataset, the Decision Tree (DT) classifier selects an average of 7 features. Similarly, in the Switzerland dataset, both the SVM and K-Nearest Neighbors (KNN) classifiers select an average of 7 features. In the Long-Beach-Va dataset, the KNN classifier also selects an average of 7 features.

On the other hand, Fig. 14.6 shows the average number of features chosen by Particle Swarm Optimization (PSO) for the same datasets and classifiers. In the Cleveland dataset, PSO selects an average of 7 features with the KNN classifier. For the Hungarian dataset, the Random Forest (RF) classifier selects an average of 7.4 features. In the Switzerland dataset, both the Naïve Bayes (NB) and KNN classifiers

Table 14.5 Performance evaluation of cleveland (UD_CL) dataset with PSO

Technique	Accuracy	Misclassification rate	TP rate/ sensitivity	TN rate/ specificity	FP rate	Precision	F-Score
PSO + DT	80.198	19.802	69.065	89.634	10.366	84.956	76.19
PSO + RF	83.828	16.172	79.856	87.195	12.805	84.091	81.919
PSO + NB	78.548	21.452	71.223	84.756	15.244	79.839	75.285
PSO + SVM	**84.488**	15.512	76.978	90.854	9.146	87.705	81.992
PSO + KNN	**84.488**	15.512	77.698	90.244	9.756	87.097	82.129
PSO + MLP	83.168	16.832	75.54	89.634	10.366	86.066	80.46

Table 14.6 Performance evaluation of hungarian (UD_HN) dataset with PSO

Technique	Accuracy	Misclassification rate	TP rate/ sensitivity	TN Rate/ specificity	FP rate	Precision	F-Score
PSO + DT	80.272	19.728	64.151	89.362	10.638	77.273	70.103
PSO + RF	81.633	18.367	67.925	89.362	10.638	78.261	72.727
PSO + NB	80.272	19.728	72.642	84.574	15.426	72.642	72.642
PSO + SVM	81.973	18.027	65.385	91.053	8.947	80	71.958
PSO + KNN	**82.993**	17.007	65.094	93.085	6.915	84.146	73.404
PSO + MLP	82.653	17.347	66.981	91.489	8.511	81.609	73.575

Table 14.7 Performance evaluation of switzerland (UD_SW) dataset with PSO

Technique	Accuracy	Misclassification rate	TP rate/ sensitivity	TN rate/ specificity	FP rate	Precision	F-Score
PSO + DT	91.057	8.943	97.391	0	100	93.333	95.319
PSO + RF	91.87	8.13	98.261	0	100	93.388	95.763
PSO + NB	76.423	23.577	81.739	0	100	92.157	86.636
PSO + SVM	93.496	6.504	100	0	100	93.496	96.639
PSO + KNN	93.496	6.504	100	0	100	93.496	96.639
PSO + MLP	92.683	7.317	99.13	0	100	93.443	96.203

select an average of 7 features. Lastly, in the Va dataset, the DT, RF, and KNN classifiers each select an average of 7 features.

Table 14.8 Performance evaluation of Long-Beach-Va(UD_Va) dataset with PSO

Technique	Accuracy	Misclassification rate	TP rate/ sensitivity	TN rate/ specificity	FP rate	Precision	F-Score
PSO + DT	73	27	88.591	27.451	72.549	78.107	83.019
PSO + RF	75	25	94.631	17.647	82.353	77.049	84.94
PSO + NB	73	27	95.973	5.882	94.118	74.869	84.118
PSO + SVM	74.5	25.5	100	0	100	74.5	85.387
PSO + KNN	77	23	99.329	11.765	88.235	76.684	86.55
PSO + MLP	75.5	24.5	98.658	7.843	92.157	75.773	85.714

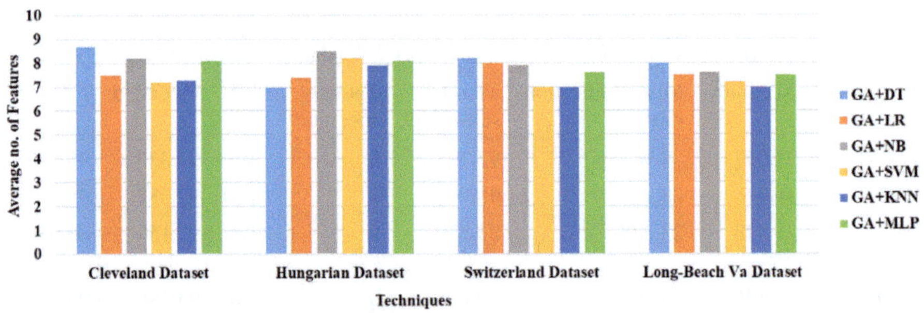

Fig. 14.5 Average number of features selected using GA

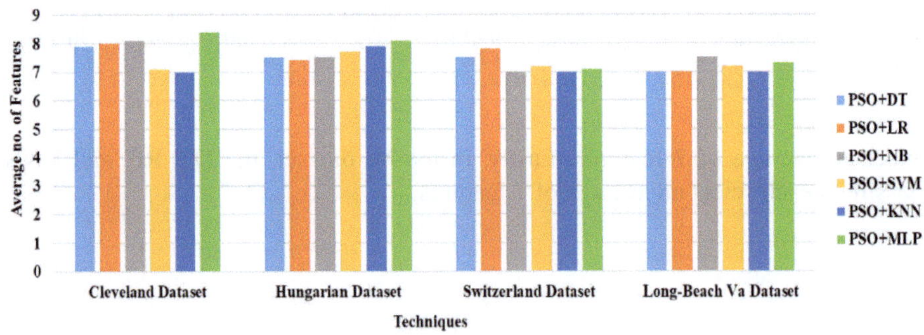

Fig. 14.6 Average number of features selected using PSO

14.4 Conclusion

The present study accentuates the need for feature selection in developing intelligent decision support systems that help diagnose a patient with heart disease. This may be achieved through metaheuristic optimization techniques such as Genetic Algorithm (GA) and Particle Swarm Optimization (PSO). Performance is thus improved through classifiers implemented over several datasets. Results reveal that PSO is performing at a better level than GA in accuracy and efficiency parameters, especially when classified within ANN and SVM classifiers. Simultaneously, the simplicity of the computation lends itself as a stronger approach from the programming standpoint to real-time applications. Results underscore the capacity of automated diagnostic systems to render medical decision-making better and reduce human error and accessibility as well as make healthcare delivery reliable. Future works could include hybrid optimization techniques and larger, more heterogeneous datasets to fine-tune systems further.

References

1. A. K. Gárate-Escamila, A. Hajjam El Hassani, and E. Andrès, "Classification models for heart disease prediction using feature selection and PCA," *Informatics in Medicine Unlocked*, vol. 19, Jan. 2020, https://doi.org/10.1016/j.imu.2020.100330.
2. S. Wadhawan and R. Maini, "A Systematic Review on Prediction Techniques for Cardiac Disease," *International Journal of Information Technologies and Systems Approach (IJITSA)*, vol. 15, no. 1, pp. 1–33, 2022.
3. M. A. Kadhim and A. M. Radhi, "Heart disease classification using optimized Machine learning algorithms," *Iraqi Journal for Computer Science and Mathematics*, vol. 4, no. 2, pp. 31–42, 2023, https://doi.org/10.52866/ijcsm.2023.02.02.004.
4. S. Wadhawan and R. Maini, "ETCD: An effective machine learning based technique for cardiac disease prediction with optimal feature subset selection," *Knowledge-Based Systems*, vol. 255, p. 109709, 2022, https://doi.org/10.1016/j.knosys.2022.109709.
5. Ş. Ay, E. Ekinci, and Z. Garip, "A comparative analysis of meta-heuristic optimization algorithms for feature selection on ML-based classification of heart-related diseases," *Journal of Supercomputing*, vol. 79, no. 11, pp. 11797–11826, 2023, https://doi.org/10.1007/s11227-023-05132-3.
6. J. Nourmohammadi-Khiarak, M.-R. Feizi-Derakhshi, F. Razeghi, S. Mazaheri, Y. Zamani-Harghalani, and R. Moosavi-Tayebi, "New hybrid method for feature selection and classification using meta-heuristic algorithm in credit risk assessment," *Iran Journal of Computer Science*, vol. 3, no. 1, pp. 1–11, 2020, https://doi.org/10.1007/s42044-019-00038-x.
7. M. Ghosh, R. Guha, I. Alam, P. Lohariwal, D. Jalan, and R. Sarkar, "Binary genetic swarm optimization: A combination of ga and pso for feature selection," *Journal of Intelligent Systems*, vol. 29, no. 1, pp. 1598–1610, 2020, https://doi.org/10.1515/jisys-2019-0062.
8. S. Taghian and M. H. Nadimi-Shahraki, "Binary Sine Cosine Algorithms for Feature Selection from Medical Data," *Advanced Computing: An International Journal*, vol. 10, no. 5, pp. 1–10, 2019, https://doi.org/10.5121/acij.2019.10501.
9. S. E. A. Ashri, M. M. El-Gayar, and E. M. El-Daydamony, "HDPF: Heart Disease Prediction Framework Based on Hybrid Classifiers and Genetic Algorithm," *IEEE Access*, vol. 9, pp. 146797–146809, 2021, https://doi.org/10.1109/ACCESS.2021.3122789.

10. G. T. Reddy, M. P. K. Reddy, K. Lakshmanna, D. S. Rajput, R. Kaluri, and G. Srivastava, "Hybrid genetic algorithm and a fuzzy logic classifier for heart disease diagnosis," *Evolutionary Intelligence*, vol. 13, no. 2, pp. 185–196, 2020, https://doi.org/10.1007/s12065-019-00327-1.
11. A. Z. Hameed, B. Ramasamy, M. A. Shahzad, and A. A. S. Bakhsh, "Efficient hybrid algorithm based on genetic with weighted fuzzy rule for developing a decision support system in prediction of heart diseases," *Journal of Supercomputing*, vol. 77, no. 9, pp. 10117–10137, 2021, https://doi.org/10.1007/s11227-021-03677-9.
12. J. Feng, Q. Wang, and N. Li, "An intelligent system for heart disease prediction using adaptive neuro-fuzzy inference systems and genetic algorithm," *Journal of Physics: Conference Series*, vol. 2010, no. 1, pp. 1–9, 2021, https://doi.org/10.1088/1742-6596/2010/1/012172.
13. B. Oluleye, A. Leisa, J. Leng, and D. Dean, "A Genetic Algorithm-Based Feature Selection," *International Journal of Electronics Communication and Computer Engineering*, vol. 5, no. 4, pp. 899–905, 2014, [Online]. Available: http://archive.ics.uci.edu/ml/datasets/Ionosphere.
14. M. T. Islam, S. R. Rafa, and M. G. Kibria, "Early Prediction of Heart Disease Using PCA and Hybrid Genetic Algorithm with k-Means," *ICCIT 2020 - 23rd International Conference on Computer and Information Technology, Proceedings*, pp. 19–21, 2020, https://doi.org/10.1109/ICCIT51783.2020.9392655.
15. A. Al Bataineh and S. Manacek, "MLP-PSO Hybrid Algorithm for Heart Disease Prediction," *Journal of Personalized Medicine*, vol. 12, no. 8, 2022, https://doi.org/10.3390/jpm12081208.
16. M. G. El-Shafiey, A. Hagag, E. S. A. El-Dahshan, and M. A. Ismail, "A hybrid GA and PSO optimized approach for heart-disease prediction based on random forest," *Multimedia Tools and Applications*, vol. 81, no. 13, pp. 18155–18179, 2022, https://doi.org/10.1007/s11042-022-12425-x.
17. E. I. Elsedimy, S. M. M. AboHashish, and F. Algarni, "New cardiovascular disease prediction approach using support vector machine and quantum-behaved particle swarm optimization," *Multimedia Tools and Applications*, vol. 83, no. 8, pp. 23901–23928, 2024, https://doi.org/10.1007/s11042-023-16194-z.
18. M. R. Ansyari, M. I. Mazdadi, F. Indriani, D. Kartini, and T. H. Saragih, "Implementation of Random Forest and Extreme Gradient Boosting in the Classification of Heart Disease using Particle Swarm Optimization Feature Selection," *Journal of Electronics, Electromedical Engineering, and Medical Informatics*, vol. 5, no. 4, pp. 250–260, 2023, https://doi.org/10.35882/jeeemi.v5i4.322.
19. J. J. Pareza Alam Jusia, Abdul Rahim, Herti Yani, "Improving Performance of KNN and C4.5 using Particle Swarm Optimization in Classification of Heart Diseases," *Resti*, vol. 6, no. 6, pp. 958–965, 2022.
20. I. Babaoglu, O. Findik, and E. Ülker, "A comparison of feature selection models utilizing binary particle swarm optimization and genetic algorithm in determining coronary artery disease using support vector machine," *Expert Systems with Applications*, vol. 37, no. 4, pp. 3177–3183, 2010, https://doi.org/10.1016/j.eswa.2009.09.064.
21. N. G. Hedeshi and M. S. Abadeh, "Coronary Artery Disease Detection Using a Fuzzy-Boosting PSO Approach," *Computational Intelligence and Neuroscience Volume*, vol. 2014, pp. 1–12, 2014, https://doi.org/10.1155/2014/783734.
22. Y. Zhang, S. Wang, P. Phillips, and G. Ji, "Binary PSO with mutation operator for feature selection using decision tree applied to spam detection," *KNOWLEDGE-BASED SYSTEMS*, vol. 16, pp. 22–31, 2014, https://doi.org/10.1016/j.knosys.2014.03.015.
23. S. Wadhawan and R. Maini, "Performance Analysis of Machine Learning Techniques for Predicting Cardiac Disease," in *2021 6th International Conference on Communication and Electronics Systems (ICCES)*, 2021, pp. 1625–1630. https://doi.org/10.1109/ICCES51350.2021.9489197.
24. G. Yang, "A modified particle swarm optimizer algorithm," *2007 8th International Conference on Electronic Measurement and Instruments, ICEMI*, pp. 2675–2679, 2007, https://doi.org/10.1109/ICEMI.2007.4350772.

Author Index

© The Editor(s) (if applicable) and The Author(s), under exclusive license to Springer Nature Switzerland AG 2026

K. Dalip et al. (eds.), *Healthcare 5.0*, https://doi.org/10.1007/978-3-032-09582-4

V

W

Y

Z